RISK ASSESSMENT IN
THE PROCESS INDUSTRIES

RISK ASSESSMENT IN THE PROCESS INDUSTRIES

Second Edition

Edited by Robin Turney and Robin Pitblado

INSTITUTION OF CHEMICAL ENGINEERS

The information of this book is given in good faith and belief in its accuracy, but does not imply the acceptance of any legal liability or responsibility whatsoever, by the Institution, the Working Party or the editors for the consequences of its use or misuse in any particular circumstances.

All rights reserved. No part of this publication may be reproduced, stored in a retrieval system, or transmitted, in any form or by any means, electronic, mechanical, photocopying, recording or otherwise, without the prior permission of the copyright owner.

Published by
Institution of Chemical Engineers,
Davis Building,
165–189 Railway Terrace,
Rugby, Warwickshire CV21 3HQ, UK.

Copyright © 1996 Institution of Chemical Engineers
A Registered Charity

ISBN 0 85295 323 2

Printed in the United Kingdom by Galliard (Printers) Ltd, Great Yarmouth.

EFCE WORKING PARTY ON LOSS PREVENTION — EDITORIAL COMMITTEE

The following people were appointed by the EFCE to undertake a revision of the ISGRA report *Risk Analysis in the Process Industries*:

R.D. Turney (Chairman)	ICI Eutech, UK
R.M. Pitblado (Editor)	DNV Technica, UK
B. Ale	RIVM/LSO, The Netherlands
A. Debeil	Debeil–Myren, Belguim
A. Opschoor	TNO, The Netherlands
G.A. Page	Cyanamide, USA

In addition, the following assisted the committee in its work by contribution to the revision:

K. Cassidy	HSE, UK
N. Hurst	HSE, UK
S.J. Kershaw	Haztech Consultants, UK
D. Leeming	HSE, UK
E. Skramstad	DNVI, Norway
P.J. Wicks	EC, Belgium
J.L. Woodward	DNV Technica, USA

ORIGINAL MEMBERS OF THE EFCE WORKING PARTY ON LOSS PREVENTION

H.J Pasman (Chairman)	TNO, The Netherlands
A.P. Cox (Secretary)	Shell, The Netherlands
E.N. Bjordal	Norsk Hydro, Norway
C.C. Brüschweiler	Sandoz, Switzerland
J. Chabanon	Rhône–Poulenc, France
J.T. Daniels	Safety and Reliability Directorate UKAEA
G. Fumarola	University of Genoa, Italy
R. Grollier Baron	IFP, France
P.L. Holden	Safety and Reliability Directorate UKAEA
P. Hyppönen	OTSO-Insurance, Finland
H.I. Joschek	BASF, Germany
E. Korjuslommi	Neste Oy, Finland
H.J.D. Lans	ICI Holland, The Netherlands
D.R.T. Lowe	ICI plc, UK
R.C. Mill	Exxon Chemical Co, USA
K. Stadel Nielsen	Dansk Sojakagefabrik, Denmark
E. Nilsson	Sprängämnesinspektionen, Sweden
G. Opschoor	TNO, The Netherlands
T. van de Putte	Directorate General of Labour, The Netherlands
J.R. Randegger	Ciba–Geigy, Switzerland
H.G. Schecke	University of Dortmund, Germany
C.H. Solomon	Essochem Europe
W.B. Howard	Consultant, USA

FOREWORD

When the International Study Group on Risk Analysis (ISGRA) (which reports to the Loss Prevention Working Party of the European federation of Chemical Engineers), which produced the first edition of this book, was set up, the topic of risk analysis was the subject of much discussion. Through its work ISGRA showed that, despite the often voiced differences, there was (and still is) a core of methodology common to all those represented by the Group. This core covered the use of thorough and systematic methods for hazard identification, scientifically based methods to assess the potential consequences of a hazard and the application of historical data, fault trees and event trees to understand the ways in which an incident may be initiated or developed. It was only in the use of quantified risk assessment (QRA), particularly by regulatory bodies, that differences in approach were significant.

In revising this book we have been able to use the thorough basis established by ISGRA. This second edition covers the main important developments which have taken place over the last ten years, particularly the improved methods available for consequence assessment and the greater understanding of the large-scale development of explosions, fires and gas dispersion.

The use of QRA is now more widely understood and accepted, although it is still the subject of much discussion. One of the most important factors leading to this change has been the adoption of the three-zone approach to criteria described in Chapter 5. The use of this approach, particularly by regulatory bodies, has dispelled much of the previous concern around 'go/no go' decisions. In addition, the extensive use of QRA offshore has demonstrated its value to industry as a tool to aid cost-effective decision making.

Since the first edition was published there has been a reinforcement of the importance of sound management systems and auditing, to support technical features and provide a continuing basis for safe operations.

As Chairman and Editor, we wish to acknowledge the efforts of members of the editorial committee and others who contributed by amending and updating individual sections of this book.

Robin Turney (Chairman)
Robin Pitblado (Editor)

CONTENTS

		PAGE
FOREWORD		v
1.	**INTRODUCTION**	1
1.1	BACKGROUND	1
1.2	THE FIRST (1985) EDITION	1
1.3	THE SECOND (1994) EDITION	2
1.4	WHAT IS RISK ASSESSMENT?	4
1.5	THE STRUCTURE OF THIS REPORT	7
1.6	DEFINITIONS	7
2.	**HAZARD IDENTIFICATION PROCEDURES**	9
2.1	INTRODUCTION TO HAZARD IDENTIFICATION	9
2.2	HAZARD IDENTIFICATION METHODS	11
2.3	ORGANIZATION OF A HAZARD IDENTIFICATION STUDY	19
2.4	REPORTING AND FOLLOW UP	23
2.5	COMPUTER AIDS AND FUTURE DEVOLOPMENTS	25
2.6	CONCLUSIONS	25
3.	**CONSEQUENCE ANALYSIS**	29
3.1	INTRODUCTION TO CONSEQUENCE ANALYSIS	29
3.2	EFFECT MODELS	31
3.3	VULNERABILITY MODELS	49
4.	**QUANTIFICATION OF EVENT PROBABILITIES AND RISK**	64
4.1	EVENT PROBABILITY ESTIMATION	64
4.2	DATA INPUTS	71
4.3	QUANTITATIVE EXPRESSIONS OF RISK	77
5.	**THE APPLICATION OF RISK ASSESSMENT**	86
5.1	SOME LIMITATIONS OF QRA	86
5.2	APPLICATION IN THE PROCESS INDUSTRY DOMAIN	87

5.3	APPLICATION IN THE PUBLIC DOMAIN	89
5.4	TOLERABILITY AND ACCEPTABILITY OF RISK	90
5.5	COST-BENEFIT ANALYSIS	93
5.6	SOME OTHER GENERAL POINTS	93
5.7	THE WAY FORWARD	95
6.	**SPECIAL TOPICS IN RISK ASSESSMENT**	**98**
6.1	OFFSHORE QRA	98
6.2	TRANSPORT RISKS	108
6.3	SAFETY MANAGEMENT SYSTEMS	113
6.4	CHEMICAL WAREHOUSE STORAGE	117
6.5	THE ENVIRONMENTAL EFFECTS OF ACCIDENTS	119
6.6	SAFETY-CRITICAL COMPUTING SYSTEMS	122
7.	**SUMMARY**	**129**
7.1	DEFINITIONS	129
7.2	HAZARD IDENTIFICATION	129
7.3	CONSEQUENCE ANALYSIS	131
7.4	QUANTIFICATION OF EVENT PROBABILITIES AND RISKS	132
7.5	THE APPLICATION OF RISK ASSESSMENT	133
INDEX		**135**

1. INTRODUCTION

1.1 BACKGROUND

In the past thirty years the rapid growth of industrial activities has resulted in many new problems related to environmental protection, energy, resource conservation and safety. Industry has been continually developing its design methods and operating techniques in order to overcome these problems. The process industries, which operate chemical, petrochemical and petroleum refining plants, handle a wide range of flammable and toxic materials which are potentially hazardous. These industries have had an excellent safety record when compared with industry as a whole. Nevertheless, the few major incidents which have occurred have made the public aware of the hazards involved. Consequently, legislation is now requiring manufacturers to demonstrate to the competent authorities that they have identified existing major accident hazards and adopted appropriate safety measures.

In this period, rapid developments were occurring in other fields of new technology, such as the use of atomic energy for power generation and in the aircraft and aerospace industries. Here, little relevant experience for assessing the safety aspects of new designs was available from the past and this led to the development of predictive quantitative risk assessment (QRA) techniques as an aid to decision making in the areas of reliability and safety.

As a result, interest in the use of QRA techniques for assessing the safety of process plant has grown considerably in Europe, both within industry and within national authorities. Much had been written and said about what is still a developing tool for safety assessment, often with subjective language, using different approaches and methodologies and giving results which are sometimes difficult to compare.

1.2 THE FIRST (1985) EDITION

The Loss Prevention Working Party of the European Federation of Chemical Engineering (EFCE) set up in September 1980 an International Study Group on Risk Analysis (ISGRA). Its terms of reference were to exchange information on QRA methods for the promotion of safety in the process industries, to increase understanding of the methods and to make recommendations on their application.

In its initial meetings ISGRA looked at the various investigative and calculation techniques which are used by the practitioners of risk analysis. As well as drawing on their own knowledge and expertise, members invited a number of workers in the field to talk on specific aspects of the subject. At an early stage in their discussions the study group members identified some areas of immediate concern in the use of quantification for the assessment of risks associated with process plants. As a result they issued a position paper 'Quantified Risk Analysis in the Process Industry'. This was initially published in *The Chemical Engineer*[1] and subsequently in a number of other European chemical engineering journals[2-5].

The members of the study group also felt that it would be useful to encourage open discussion of their provisional views before writing their report. Therefore they proposed to the EFCE Loss Prevention Working Party that they should be allocated a session at the 4th International Symposium on Loss Prevention and Safety Promotion in the Process Industries, in Harrogate, UK, in September 1983.

This proposal was accepted and the session at the symposium[6] led to some lively discussion. This helped the study group members crystallize their views on some of the more contentious aspects of the use of QRA. The members represented a wide variety of organizations which are interested in use of QRA as a tool to assist with making decisions on the safety of process plant. Some used it regularly, whilst others had strong reservations about its use. The opinions of these members could not always be reconciled and so, although the majority of the first report was supported by all the members, in some places the views expressed represented a majority rather than a unanimous opinion.

1.3 THE SECOND (1996) EDITION

In 1993 the EFCE Loss Prevention Working Party under the chairmanship of H. J. Pasman, appointed an editorial committee to undertake the revision of this book. Members of the committee and other contributors are listed on page iii.

In carrying out the revision, the editorial committee was faced with a very different task to that of the original group. Many of the techniques described in the book are now widely accepted by the process industries throughout Europe, methods have been improved and new fields of application developed. At the same time the use of QRA still raises concern in a number of quarters.

The basic approaches to hazard identification are now generally accepted and widely used. There is a trend towards the use of hazard studies from the earliest stages of a project. At this point the opportunities to incorporate simpler inherently safe methods for the elimination, reduction or control of hazards are at their greatest. Computer systems are being used as an aid in the recording of studies but more ambitious applications for hazard identification remain at the trial stage.

In consequence assessment (Chapter 3) important research work has been carried out in the fields of dispersion, vapour cloud explosion and fire radiation, with the use of large-scale experimental facilities. These have lead to fundamental reassessment of the models used for vapour cloud explosion with improved understanding of the circumstances necessary to create an explosion. Experimental work has also improved the understanding of the mechanisms involved when flashing liquids are released with the formation of droplets and aerosols together with the associated rain-out and pool formation. Large-scale experiments in gas dispersion, which were carried out in the early 1980s, have been used to refine the computer models used in this area and a number of important comparative studies have been made. Since the first edition the computer models have also changed from predominantly single consequence models, to multi-purpose tools able to carry out a number of linked calculations, often with the inclusion of data banks of selected physical properties and graphical outputs. These tools are also much more 'user friendly'.

The basic methods used in the quantification of event probabilities and risk have changed little, since the first edition of this book. One of the greatest problems that remains is the selection of failure rates which adequately represent the particular problem being studied. Data collection exercises have been carried out in a few well-defined areas, notably offshore, but limited work has been carried out elsewhere. A bench marking exercise on QRA carried out with the support of the Commission of the European Communities, showed that the results obtained by different assessors can vary widely.

Overall, the use of QRA has continued to increase, although the pattern is by no means uniform across Europe and some countries still remain sceptical of the value of the technique.

In other countries, moves away from prescriptive forms of regulation towards 'goal setting', in which industry is required to demonstrate that a particular facility is capable of meeting defined goals, has given positive encouragement to the use of QRA. Important studies, based initially on public enquiries into nuclear power, have improved understanding of the way in which risk assessment may be used. A two-bound approach, with an upper bound above which a risk is considered intolerable and a lower bound below which it is broadly acceptable, is now well established. Between these two, the principle of ALARP (as low as reasonably practicable) is applied. Concepts of societal risk are now seen as being highly dependent on the overall societal/economic value of a facility and there is a widespread appreciation of the need for great care in the application of fixed criteria.

A number of new uses of risk assessment have developed since the first edition. The Piper Alpha and Alexander Kielland tragedies led to the introduction of regulations requiring the use of QRA in both the UK and Norwegian sectors of the North Sea. QRA is also being applied to assess the transport of hazardous substances, the risks associated with environmental incidents and, to a limited extent, fires in chemical warehouses. Brief résumés of work in these fields, together with the developments in the assessment of management systems and safety-critical computer systems, have been included in this edition for the first time.

1.4 WHAT IS RISK ASSESSMENT?

When used for the assessment of hazards in process plant and storage, risk analysis sets out to answer three questions:
- What can go wrong?
- What are the consequences and effects and are these acceptable?
- Are the safeguards and controls adequate to render the risk acceptable?

The first step of hazard identification (what can go wrong) is purely qualitative and is often called a hazard or safety study. Such a study may reveal aspects of the facility which require more consideration. An important step is then to analyse the effect or consequences. In some cases this might be quite simple, making use of qualitative measures; in other cases quantitative methods described in this book may be required. The assessment of consequence has two uses: first it enables an assessment to be made about whether these are acceptable and second the results can be used as a basis for emergency planning. The principles of inherent safety can be applied to either reduce or eliminate undesired consequences.

If the consequences are not acceptable and their magnitude cannot be reduced, consideration needs to be given to the incorporation of control measures which will prevent an incident occurring. In some cases, codes of practice and standards developed by industry may provide sufficient evidence that, if they are properly applied, the resultant risk is acceptable. In other cases, where the consequences are very serious, it is likely that a quantitative estimation of the event probability and risk must be made. These are the cases covered in this book.

The resultant estimate of the risk may then be compared with agreed criteria. If the risk meets the criteria it can be classed as broadly acceptable. If it does not, it will be necessary to make improvements either to reduce the likelihood of the event or to reduce the consequence further. In an extreme case

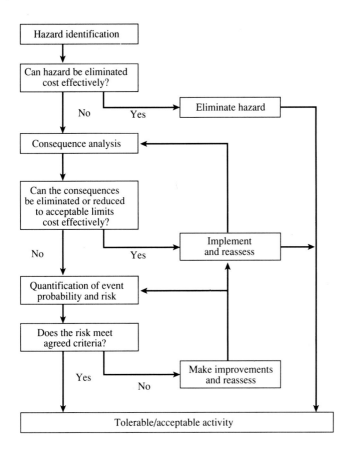

Figure 1.1   Procedure for the application of risk assessment.

where such improvements cannot be made, the risk may be classed as intolerable. Figure 1.1 shows the procedure.

Many companies do not use quantitative techniques after the identification stage. However, decisions are made and actions taken to control specific hazards, and they are done considering probabilities and consequences qualitatively. In a sense, this is an elementary form of risk analysis, but at a less sophisticated level than assessments involving quantitative consideration of probabilities and consequences. The study group did not examine in depth the way in which these qualitative assessments are made because its terms of reference related primarily to a study of the use of QRA.

The major part of this book is devoted to the techniques and methods by which risk analysis is carried out. It is, however, important to remember the wider issues associated with risk analysis. As well as the topic of criteria, discussed in Chapter 5, the wider issues include individual and professional responsibility. Bjordal[7] notes the need for those carrying out risk assessments to remain aware of the basic reason for their work and to ask 'What are the conditions that must be established for safe operation?', as well as seeking safer alternatives. Bjordal also notes the importance of the analyst being honest in the way data is used and presented, as well as in the overall way in which the task is organized.

These issues have been developed more fully by the UK Engineering Council in a code of practice prepared with the support of the 44 UK engineering institutions represented by the Council. The 10-point code *Engineers and Risk Issues*[8] is summarized in Table 1.1. A series of case studies issued in conjunction with the code show that issues of hazard identification, risk control and risk communication are important in all branches of engineering. The code and its supporting guidelines emphasize that engineers, by their involvement and understanding, have a central role in the control of risk, as well as with responsibilities to exercise reasonable professional skills and care in the performance of their work. Whilst recognizing that most risk issues can be resolved at the local level, the code sets out an approach which can be followed when an engineer considers that an unacceptable level of risk is being allowed to persist. Through its links with the professional and educational institutions the code has made an important addition to thinking on risk issues and the understanding of them.

Table 1.1
The 10 Point Code of Professional Practice on Risk Issues .*

1. **Professional responsibility**	Exercise reasonable professional skill and care
2. **Law**	Know about and comply with the law
3. **Conduct**	Act in accordance with the codes of conduct
4. **Approach**	Take a systematic approach to risk issues
5. **Judgement**	Use professional judgement and experience
6. **Communication**	Communicate within your organization
7. **Management**	Contribute effectively to corporate risk management
8. **Evaluation**	Assess the risk implications of alternatives
9. **Professional development**	Keep up to date by seeking education and training
10. **Public awareness**	Encourage public understanding of risk issues

* Courtesy of the Engineering Council.

There are important ethical issues raised in decision-making regarding public risks. The Royal Society has sponsored two symposia to address some of these issues[9,10].

In general, many companies in Europe are actively trying out risk-based decision-making. They are doing so because it is effective in the context of goal-setting legislation, which is being used increasingly by safety authorities. Few engineers express serious problems with the component steps in risk analysis (Figure 1.1, page 5), except the final stage where a total risk estimate is generated and compared to a criterion. This stage continues to be the subject of significant industry debate.

1.5   THE STRUCTURE OF THIS REPORT

In preparing the second edition of this book, the structure of the first edition has been retained. The main part of it covers the process of risk assessment from hazard identification, through consequence analysis and the quantification of event probabilities and risk, to the comparison of the calculated risk with appropriate criteria.

Chapter 2 has been reorganized to improve readability. Appendices on consequence analysis which appeared in the original edition have been incorporated into Chapter 3. Considerable advances have been made in the field of explosion analysis and these, and the remaining difficulties, are reviewed in this edition of the book. In other sections original material has been maintained wherever relevant. References have been reviewed, new ones included and with preference given to review papers.

Chapter 6 attempts to provide a series of brief overviews of those areas which are still developing. It needs to be stressed that some of these techniques are still at the development stage and are not all broadly supported at the present time.

1.6   DEFINITIONS

One of the first matters which the study group addressed was the definition of the terms used in risk analysis. Various publications[11,12] have defined words such as hazard and risk in different ways. Therefore it was decided that ISGRA would not formally propose definitions for the more commonly used words and terms. Rather it would use them in a uniform manner and in such a way that their meaning is clear.

REFERENCES IN CHAPTER 1
1. ISGRA, 1982, Quantified risk analysis in the process industries, *The Chemical Engineer*, 385, 385–386, 389.
2. *Ingegneria Chimie Italia*, 1983, 19 (1–2), GEN, 17 (in English).
3. *Chemie Ingenieur Technik*, 1983, 55 (9), A404 (in German).
4. *Informations Chimie*, 1983, 236 (in French).
5. *Kemisti Kemisten*, 1984, 11 (2), 123 (in Finnish).
6. 4th International Symposium on Loss Prevention and Safety Promotion in the Process Industries. Harrogate 12–16 Sept. 1983, Session G: risk assessment, *IChemE Symposium Series No. 80*, (IChemE), G1–G55.
7. Bjordal, E., 1991, Risk as a basis for our thinking and decisions about the future, *Risk 2000, London 1993* (IBC).
8. *Engineers and Risk Issues — A Code of Professional Practice and Guidelines on Risk Issues*, 1993 (Engineering Council, London).
9. *The Assessment and Perception of Risk*, 1980 (The Royal Society).
10. *Risk:Analysis, Perception, Management*, 1992 (The Royal Society).
11. Methodologies for hazard analysis and risk assessment in the petroleum refining and storage industry, 1982, *report no. 10/82* (CONCAWE).
12. Jones, D.A., 1992, *Nomenclature for Hazard and Risk Assessment in the Process Industries*, second edition (IChemE).

2. HAZARD IDENTIFICATION PROCEDURES

This chapter aims to promote a better understanding of the application and limitations of methods for identifying hazards in the process industries. From examination of published methods for hazard identification, together with those used by the various companies and organizations represented in the EFCE Loss Prevention Working Party, the editorial committee found that no single method can be recommended for all circumstances. By exchanging their experiences of using hazard identification, however, the members were able to offer advice on how to select a hazard identification method to suit both the needs of the process and the experience of the organization applying the methods.

2.1 INTRODUCTION TO HAZARD IDENTIFICATION

2.1.1 THE NEED FOR HAZARD IDENTIFICATION

Many people believe that hazard identification is the most important step in risk analysis on the grounds that 'a hazard identified is a hazard controlled'. In support of this statement they observe that a reputable company is unlikely to expose its business to the financial consequences which could result from failing to eliminate or control a major hazard. In cases where losses have occurred, the cause is often the failure of the organization or the individuals to use their knowledge, rather than lack of knowledge about how to prevent the incident. It should be recognized that even a responsible organization may have major quantities of hazardous materials within its jurisdiction, without necessarily having the expertise to identify and analyse the hazards and potential consequences of mishaps. This has been known to occur where the hazardous materials are ancillary to the organization's main activities — for example, the bulk storage of liquefied petroleum gas (LPG) at a brickworks or of chlorine at a water treatment plant. In both these cases the organizations may not have the experience or expertise to carry out realistic assessments of the hazards associated with multi-tonne quantities of LPG or chlorine.

While not everyone will agree that 'a hazard identified is a hazard controlled', there is no doubt that hazard identification methods make a very important contribution to plant safety by helping an organization either to apply its knowledge systematically, or to seek outside help where the hazard is beyond its own experience and knowledge. Without highly effective methods for hazard identification and control, the process industries could not have reached their present stage of development.

Discussions which followed the publication of many of the methods described in this chapter gave the impression that hazard identification is done only on the process design flow sheets, and there is a danger that this chapter will give further support to that impression. It must be emphasized that hazard identification begins with the concept of the project and continues in different forms throughout the life of the plant.

A hazard identification study on a completed process design is important for two reasons. First, it is a check that the organization's knowledge of hazard control has been properly applied during the process development and design stages of the project. Second, the hazard identification study report provides the foundation for the plant 'safe operating procedures' which will be used daily until the plant is scrapped — or until it is modified, when new ones should be prepared.

It is for these reasons that hazard identification at the design stage of a project has received and will continue to receive more attention than at any other stage of a project. The application of hazard identification methods at other stages must not be neglected, however, particularly when plant modifications are made.

2.1.2 APPLICATION OF HAZARD IDENTIFICATION METHODS TO NEW DESIGNS AND EXISTING PLANTS

The application of hazard identification methods can be both time consuming and expensive, and thus some sections of a plant will receive more detailed study than others. The depth of the study depends upon an appraisal of the inherent hazards in the various sections of the plant. (see 'Hazard Indices', page 11).

In the case of a highly sensitive reactor system, the hazard identification study may be very detailed and often supplemented by a reliability analysis of the control system using a method such as fault tree analysis. On the other hand, a service unit plant might only be reviewed for operability and personnel protection. Therefore, the depth and scope of a study is determined by an organization's perception of the hazards in a process and its appraisal of the need to control them. Regulations defining major hazard inventories for a wide range of materials also affect decisions on scope.

In stressing the need to identify hazards as early as possible in the development of a process, and especially at the process design stage, the reader may have formed the impression that hazard identification ends when the design specification has been approved. In fact approval of a design means only that 'at the time of the study, the study team believe that — provided the plant is constructed and operated in accordance with their recommendations — the plant will be acceptably safe'. The first uncontrolled change during construction or the first unapproved modification during operation potentially negates this approval. It is essential that a management-of-change control procedure is in place to take care of these situations.

Regulations in a number of European countries as well as the USA now require the application of hazard identification to existing plants presenting major hazards. Responsible operators have recognized the value of this and many now apply hazard identification to all existing plants, even where this is not required by legislation[1].

People who have applied hazard identification procedures to existing plants claim that participants in a study team gain a much broader understanding of their jobs than could be obtained by any other training method. They have noticed that the resulting improvement in employee motivation is reflected by an overall improvement in the efficiency and safety of the operation.

2.2   HAZARD IDENTIFICATION METHODS

2.2.1   HAZARD IDENTIFICATION TECHNIQUES IN THE PLANNING STAGE

The degree to which it is economic to eliminate a hazard is very dependent upon the stage of the project when the hazard is first recognized. For example, if a hazard is discovered at the pilot plant or process development stage, it might be possible to eliminate the hazard entirely by selecting a safer process route. By the time the design has reached the stage of being sufficiently documented to allow a detailed hazard identification study, the flexibility to eliminate hazards entirely is very much reduced. Therefore, it is often beneficial to carry out the hazard identification in stages matched to the quality of information available, particularly where a significant element of new technology is present in a project. For example, as soon as the process raw materials and intermediate products are known, a short hazard identification review should be carried out to identify any areas where more knowledge about hazardous properties of the materials is needed. Research can then be initiated to provide the knowledge to help select the most inherently safe process route, as suggested by Kletz[2].

Hazard indices
Hazard indices such as those in Dow's *Fire and Explosion Index*[3], *The Mond Index*[4], and the check-lists of DGA[5] are designed to give a quantitative indication of

the potential for hazardous incidents associated with a given design of plant. These methods are particularly useful in the early stages of hazard assessment in that they require a minimum of process and design data and can graphically demonstrate which areas within the plant require more detailed attention. They can also help to identify which of several competing process routes will present the least hazards.

The indices can also be used to identify which installations require hazard identification and which methods to use. Index techniques do not generate specific accident sequences for further analysis.

2.2.2 HAZARD IDENTIFICATION TECHNIQUES IN THE DESIGN STAGE

Hazard identification has long been an integral part of design and operational practice. In the past, however, it was often an informal process dependent on how well company experience was documented and applied to subsequent designs. At times it depended on the experience of those directly involved. The identification methods discussed in this section are generally formal procedures which are carefully structured to improve the completeness of hazard identification.

Hazard identification is aimed at two particular outcomes. First, there is the identification of serious incidents which may result directly in danger to employees or the public, or in financial loss. These are usually known as the 'top events'. Second, the fundamental methods can be used to identify the underlying root causes which can lead to the top events, as well as to identify those incidents which could lead to operability, maintainability and other problems.

The techniques most frequently used to identify hazards in the design stage can be grouped as follows:
- Basic techniques:
— Hazop
— What-if
- Derivations of basic techniques:
— Knowledge based Hazop
- Supplementary techniques:
— Check-lists
— Failure mode and effect analysis
— Fault tree analysis
— Event trees
— Task analysis

Hazard identification procedures involve a systematic review of the equipment on each drawing. In earlier reviews this will usually be on a 'plant

HAZARD IDENTIFICATION PROCEDURES

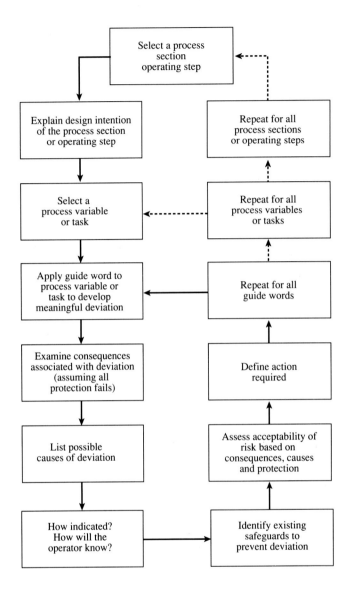

Figure 2.1   Hazop analysis method flow diagram

item' by 'plant item' basis, while on the completed design specifications it will usually be done in more detail 'line by line'. The methods differ in the techniques used to reveal the potential hazards and in the degree of detail.

For process installations, the techniques mostly recommended and used are the Hazop and 'What-if' methods. These methods use a team of people

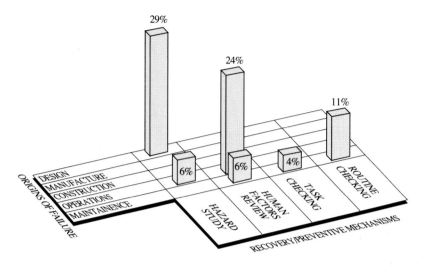

Figure 2.2 Tower blocks sum of direct causes classified by origins and prevention/recovery mechanisms.

with individual skills and experience which are relevant to the type of process and equipment to be reviewed.

Hazop : guide-word method
The most widely used technique is the hazard and operability (Hazop) method developed by ICI, published by Lawley and well described by the Chemical Industries Association[6] and by the Center for Chemical Process Safety (CCPS)[7]. The method uses guide words such as 'more', 'less' and 'reverse' which can be applied to the process parameters to generate deviations from the 'designer's intention'. This produces questions like 'What if there is "more flow" than design?' 'What if there is reverse flow'? In its published form it is completely general in nature and can be applied to any process be it batch or continuous, existing or new project or procedures. A diagram derived from the CCPS document shows a flow diagram for Hazop analysis (Figure 2.1, page 13).

Although simple in concept, the Hazop method has proved to be very effective over a number of years. Hazop is especially effective when applied to new or novel plants and processes.

A recent study (Bellamy *et al*, see Chapter 6, Reference 12) of past incidents indicated that the use of Hazop would, by itself, probably have prevented 29% of the design incidents and 6% of the operational incidents a higher proportion than any other single technique (see Figure 2.2).

Table 2.1
A list of guide words.*

Guide words	Meanings	Comments
NO or NOT	The completion negation of these intentions.	No part of the intentions is achieved but nothing else happens.
MORE LESS	Quanititative increases or quantitative decreases.	These refer to Quantities + Properties such as flow rates and temperatures as well as Activities like 'HEAT' and 'REACT'.
AS WELL AS	A qualitative increase.	All the design and operating intentions are achieved together with some additional activity.
PART OF	A qualitative decrease.	Only some of the intentions are achieved; some are not.
REVERSE	The logical opposite of the intention.	This is mostly applicable to activities — for example, reverse flow or chemical reaction. It can also be applied to a substance — for example, 'POISON' instead of 'ANTIDOTE' or 'D' instead of 'L' optical isomers.
OTHER THAN	Complete substitution.	No part of the original intention is achieved. Something quite different happens.

* Courtesy of the Chemical Industries Association.

As with any of the techniques listed here, Hazop will not by itself ensure a safe plant. It needs to be supported by well-defined engineering standards and a comprehensive safety management system.

The generality of this method can lead to difficulties in application, particularly for less experienced teams, and it is often helpful to produce a more specific set of guide words or phrases to prompt the questions, appropriate to the types of process being considered. For example, guide words may be produced for continuous processes, electrical power systems, assessing operating instructions, etc.

Table 2.1 lists the classic set of guide words, often used for the analysis of continuous and batch processes as well as of operating procedures[6]. Some companies utilize, especially for continuous processes, a set of 'guide phrases' in which they associate a parameter to each guide word (see Table 2.2, page 16).

Table 2.2
Extended hazop guide words for continuous chemical processes.

NO FLOW	MORE TEMPERATURE	SAMPLING
REVERSE FLOW	LESS TEMPERATURE	CORROSION/EROSION
MORE FLOW	MORE VISCOSITY	SERVICE FAILURE
LESS FLOW	LESS VISCOSITY	ABNORMAL OPERATION
MORE LEVEL	COMPOSITION CHANGE	MAINTENANCE
LESS LEVEL	CONTAMINATION	IGNITION
MORE PRESSURE	RELIEF	SPARE EQUIPMENT
LESS PRESSURE	INSTRUMENTATION	SAFETY

Knowledge-based Hazop
A variant of the guide word Hazop is the knowledge-based approach[7,8] in which guide words are supplemented or partially replaced by both the company's and team's knowledge supported by specific check-lists.

The knowledge base is used to compare the design to well established basic design practices that have been developed and documented from previous plant experience. The implicit premise of this version of the Hazop analysis technique is that the organization has extensive design standards that the team members are familiar with. An important advantage of this method is that the lessons learned over many years of experience are incorporated into the company's practices and are available for use at all stages in the plant's design and construction. Thus the Knowledge-based Hazop study can help ensure that the company's practices and therefore its experience, have indeed been incorporated in the design. It needs to be recognized that few companies, other than industry leaders, have the detailed engineering design standards and expertise necessary to support this approach.

Sometimes a check-list can be used to help a team develop deviations for a particular study. Appendix B of Reference 7 contains a comprehensive check-list that leaders can use to supplement their knowledge when leading a Hazop team.

In the knowledge-based Hazop method, the guide-word Hazop approach is used to supplement the check-list approach to ensure that new problems are not overlooked when portions of the process involve major changes in equipment technology or novel chemistry. Particular attention needs to be paid to sources of hazard between sections of plant and between the plant and offsite areas.

"What-if and What-if/check-list methods
The 'What-if and What-if/check-list' methods[7] are creative, brainstorming examinations of a process or operating procedure, similar to the Hazop method. They are carried out in a small team with a chairman asking questions. The questions and answers are recorded.

The 'What-if' analysis considers the results of unexpected events that would produce an adverse consequence. It involves examining deviations from the design by asking 'What-if' questions. The experience of the chairman is of crucial importance. The chairman normally prepares most of the questions ahead of the sessions. Check-lists focusing on the types and sources of hazards associated with the process are useful to make sure the analysis is complete. The combination of the creative 'What-if' method and the experience-based thoroughness of a check-list provides a powerful hazards identification technique.

A 'What if analysis' usually reviews significant sections of a process simultaneously, rather than adopting a line-by-line approach. The results of a 'What if' analysis address potential accident situations implied by the questions and issues posed by the team. Examples of 'What if' questions are:
- What if the raw material is the wrong concentration?
- What if cooling is lost?
- What if the vessel agitation stops?

This leads to faster execution times than Hazop, although the depth of coverage is less thorough, and greater opportunity exists for missing subtle hazards unless the check-list prompts for these. The technique is commonly used for batch operations, and in the pharmaceutical sector.

Check-lists
Check-lists specify those aspects of a plant which require attention for safe design. The lists are derived from industry codes, regulations, past accidents and judgement. They are particularly helpful in ensuring designers address known hazards, but check-lists are *not* effective on their own to address new hazards or novel technology.

Failure modes and effect analysis
Failure modes and effect analysis (FMEA)[7] evaluates the ways equipment can fail (or be improperly operated) and the effects these failures can have on an installation. These failure descriptions provide analysts with a basis for determining where changes can be made to improve a system design. During FMEA, hazard analysts define single equipment failures (usually electrical or mechanical) and determine the effects both locally and on the whole system. Each individual

failure is considered as an independent occurrence with no relation to other failures in the system, except for the subsequent effects that it might produce.

FMEA is most suited to installations where the danger comes from the mechanical equipment, electrical failures etc, but not the whole process (and contrasts with Hazop which is applied to whole processes, where the danger comes from hazardous materials in chemical process systems). FMEA can be used to supplement Hazop for specific equipment, such as package units, but it is *not* recommended as a general identification method for chemical processes.

Fault tree analysis
A fault tree analysis (FTA)[7] is a graphical model that illustrates combinations of failures that will cause one specific failure of interest, called a 'top event' — such as 'explosion in reactor'. The FTA process, in addition to identifying the root causes of top events, sometimes reveals alternative outcomes of those root causes or common causes of failure. In this way the method can be used qualitatively to help identify further top events. Chapter 4 deals with the quantitative use of FTA in QRA (page 67). The technique of 'minimum cut sets' may be also used in conjunction with FTA to determine the minimum set of coincident failures necessary for the hazardous event to occur.

Both FMEA and FTA are useful aids to hazard identification as they both structure and document the analysis. But, because they involve very detailed analysis of components and operations, their use in the process industry is mainly limited to identification of special hazards where they form the basis of quantification of risks.

Event tree analysis
Event tree analysis (ETA)[7] evaluates the potential for an accident as the result of a general equipment failure or process upset, known as an initiating event. Unlike FTA (a deductive reasoning process), ETA is an inductive process where the analyst begins with an initiating event and develops the possible sequences of events that lead to potential accidents. Event trees provide a systematic way of recording the accident sequences and defining the relationships between the initiating events and subsequent events that result in accidents. In this manner, it can play a useful role in hazard identification. ETA is a critical part of most QRA studies (see Chapter 4). An example of an event tree is given on page 70.

Task analysis
Reliable human performance is necessary for the success of human/machine systems and is influenced by many factors, such as stress, emotional state, training

and experience, or work environment and hardware interfaces. Many of the hazard identification techniques described in this Chapter can be used to pinpoint the potential for accidents caused by human failure. For an example, typical Hazop analyses frequently include general operator errors as causes of process deviations.

Task analysis[9] is used to analyse the human characteristics of systems, operations and procedures to identify likely sources of error. The use of task analysis is generally limited to situations where other techniques such as Hazop, What-if or FTA have shown that human errors could lead to high risk.

2.3 ORGANIZATION OF A HAZARD IDENTIFICATION STUDY

2.3.1 GENERAL OBJECTIVE

The main objective of a hazard identification study is to identify hazards and decide on which actions to recommend, but not to solve problems or redesign during the sessions. When a hazard has been identified a decision must be taken and recorded about what action is considered appropriate. It may be that the consequences are so small, or the chance of the unwanted event happening so remote, that no change is necessary. On the other hand, it may be that some change of process or detailed design is needed to eliminate or reduce the risk from this source.

Often the decision will be obvious on purely qualitative evidence. Occasionally, however, the appropriate action will not be obvious, particularly where the potential consequences are severe. In such cases QRA techniques can be used to put the hazard in perspective and improve confidence that proposed improvements are effective and that their cost is consistent with their benefit. The QRA techniques are discussed in more detail in Chapters 3 and 4. In many cases insufficient information is available within the meeting to make a decision and it is necessary to detail someone to study the problem outside the meeting.

The end product of a structured hazard identification study is normally a report from the hazard study team. The hazard identification report lists safeguards and recommended actions, as concluded by the team, pointing out hazards which — in the opinion of the study team — may not be adequately controlled in the proposed or existing design.

2.3.2 RESPONSIBILITY OF THE TEAM

It must be appreciated that a hazard identification study cannot simply be grafted on to an existing set of management and decision-taking procedures. It should be properly integrated with other procedures and the responsibilities for resultant actions clearly understood. In particular it must be clear that the team's remit is to identify the hazards and not to be involved in the development and modification of the design in response to the hazard identified. The latter approach creates two major problems. First, there is a temptation to resolve debate over minor additions with the argument 'it doesn't cost much — put it in'. This can eventually lead to unnecessary proliferation, particularly in instrumentation and valves, which adds considerably to both complexity and cost. Second, there is a danger that an 'instant design change' thought up in the meeting will be owned so strongly by the team that it may fail to examine the change as deeply as the original design. It is not unknown for such a change to eliminate the first hazard but at the same time introduce another, perhaps more serious, problem.

Thus on balance it is recommended that the team should concentrate on identifying hazards and deciding which should be examined further. These should then be passed on via some formalized documentation, usually the hazard identification report from the team, for decision-making by the management for possible further study and redesign.

In some companies, hazard identification is carried out as part of the design development, and decisions on design improvements are taken in the study meetings. If this approach is adopted it is essential that the appropriate level of authority is present at the meeting and that the process is carefully controlled to prevent the natural 'ownership' of the design from inhibiting a thorough objective examination.

2.3.3 COMPOSITION OF THE TEAM

There is no hard and fast rule for the composition of a hazard identification team, but there are some basic guidelines[6,7]. It should include people with experience of design and operation of the unit operations which make up the process and experience of the raw materials and products involved. To be effective the team must be interactive and allow each person to make a contribution. Therefore, the knowledge and experience should be concentrated in a small number of people, typically four to six. Ideally, the team members should be sufficiently senior to be accountable for decisions taken in the meetings, and at least one member of the study team should be independent of the project. A good leader with experience and training of hazard identification is needed to resolve debate, bring out conclusions and ensure constructive progress via the systematic approach.

A typical study team might be:
- independent study team leader;
- recorder/scribe (optional);
- design/process engineer responsible for plant section being studied;

- operating manager with relevant experience, ideally the person who will be responsible for commissioning the plant section being studied;
- instrumentation/control engineer;
- safety/environmental/industrial hygiene specialist (where specific expertise is required);
- other specialists such as electrical, maintenance, machinery, furnace engineers and research chemists (as required).

2.3.4 PREPARATIONS FOR THE STUDY

The success of a hazard identification study depends upon the quality of the documentation which the team is asked to review and the experience of the team members. In addition, it is essential that the responsibility of the team and its interface with other aspects of the project or operating management be fully understood.

The hazard identification study team leader prepares a package of information containing the documentation discussed later in this section and makes it available to the team members.

Main points of the initiation of the study:
- The leader states objectives and boundaries of the study.
- The leader proposes the hazard identification method to use, programme, schedule and participants.
- The designer presents an overview of the process and most important feature of the design as follows:

— An overview of the process giving the throughput of the main unit and typical operating conditions.

— A consideration of whether the process is new to the company or the company's experience with similar plants. The design team's experience of the type of process or similar processes.

— The areas where special hazards have already been identified by the design team and previous hazard reviews, and what — if any — changes have been made to eliminate or control these hazards.

- The leader presents accident/near misses history of the process.
- The team members discuss major hazards, based on historical data and their own experience.
- The proposed location and siting relative to other plants and a preliminary equipment layout.

Process drawings

The process drawings used must truly represent the up-to-date design of the project or the existing plant. The detail of information required obviously depends on the objectives of the study and the phase of the project. For the most detailed study, often used to finalize the piping and instrumentation diagrams (P&IDs), the design package should provide information on such factors as process piping ratings, materials of construction and other details which are essential to enable the safety of the design to be assessed. Before the design is started a standard format for the information to be provided in the design package and on the process flow sheets should be agreed.

Sources of knowledge

Knowledge of hazards is acquired in one or more of the following ways:
- by personal experience of the hazard;
- by consulting someone else who has experience of the hazard;
- by realizing that a given situation is similar to a hazardous situation of which you are aware;
- through the knowledge accumulated in engineering codes and practices;
- accidents and 'near misses' history (see next section).

Individual experience, broadened by information about the experiences of others, is the fundamental ingredient of hazard identification. However, hazard control techniques require that individual experiences are collected and recorded for posterity in a form which makes the knowledge readily accessible to the people developing and designing the equipment. National and international codes and standards are examples of this. The authors of a code are implying that, 'based on our collective experience, equipment designed to this code will be acceptably safe'. Consequently, codes and practices provide minimum standards (norms) against which deviations from safe practices can be identified and appraised. Systematic comparison of a design specification with recognized codes and practices forms the basis of knowledge-based hazard identification methods.

In cases where portions of new processes lie outside the scope of existing codes or practices, prior experience must be supplemented by information obtained by the research and development engineers, together with the personal knowledge of the hazard identification team members. Consequently, the hazard identification methods used in such cases have to be directed towards stimulating the team members to use their own experience of 'safe' and 'unsafe' as the standard against which to appraise the design. These techniques, referred to as 'basic techniques', are essentially structured ways for stimulating a group of people to apply their knowledge to the task of identifying hazards mainly by raising a series of 'What-if?' questions. They have the advantage that they can be used in any industry, whether or not codes of practice are available.

Accidents and 'near misses' history
Industrial history contains many cases where the knowledge to prevent a loss existed but, in the absence of an organization to identify and control hazards, the knowledge was not available to the people who could have avoided the incident1[10].

In order to help identify possible hazards for the process to be reviewed, information about accidents that have happened and from 'near misses' should be collected from inside and outside the company. Examples of external databases are: MHIDAS (AEA Consultancy Services, UK), FACTS (TNO, Netherlands), Loss Prevention Bulletin Index (IChemE, UK), WOAD (DNV, Norway).

Data on materials processed and equipment
Identification of potential hazards requires information on the materials being processed as well as on the equipment used in the process.

It is essential that the availability of this information is checked at an early stage of the project, ideally as soon as a proposed process is available and before the flow sheet is firmly established. For well-tried processes and common materials, this data will be readily available when required. However, if new conditions, chemicals and materials are involved it may take several months to design and carry out experiments to produce the data.

The pertinent data required for each study includes the following[11]:
• physical and chemical properties of the materials in the process;
• toxicological and biological properties (biodegradability, bio-accumulation, persistence, etc);
• reaction parameters (enthalpy, onset of exothermal reactions, etc);
• thermal stabilities, including properties of decomposition products;
• reactivity of process chemicals with materials of construction;
• flammability and explosion limits for fuel/oxidant/inert gas mixtures, including dusts;

Structured procedures have been developed to ensure that this, and other information necessary for an effective study, is collected and reviewed from the earliest stages of a project[12].

2.4 REPORTING AND FOLLOW UP

The record of the hazard identification study should be designed to *'meet the needs of the user's organization'*. Hazard identification is primarily an action-oriented procedure and the prime objective of the documentation is to ensure that appropriate actions are taken and recorded for future use.

It is important when defining the documentation needed, to consider the total life of the plant and not just the project phase. Good records can help:
- in clarifying the role of safeguards and defining the envelope of safe operation;
- in the preparation of operating and emergency procedures;
- ensuring that, when modifications are undertaken, those making the change are aware of the factors considered by the hazard identification team;
- as a basis for auditing the operation of the plant.

More and more industries are required by legislation to prove that they have identified all the hazards of their installations, new as well as existing. This makes it necessary to keep complete records of hazard identification studies to prove, on request from authorities, the thoroughness of their studies. (This applies to Hazop as well as to FMEA, What-if and other hazard identification techniques). Current good practice recognizes the importance of recording all significant hazards identified and the preventative and corrective measures (safeguards) incorporated into the design.

Complete records should include:
- a log of meetings held;
- names of those that have attended;
- complete worksheets together with a set of drawings marked up;
- records of potential problems found, possible consequences, safeguards, recommended actions and actions taken.

For ease of reference each action recommended by the team should be numbered sequentially. Future correspondence and replies from the design team should also refer to the recommended action number.

It is important that someone is given responsibility to ensure that all the actions raised by the team are completed. That person should review the designer's written answers to the hazard identification team's report and where appropriate, contact the other team members before accepting the proposed solutions.

Figure 2.3 (pages 26–27) is an example of a Hazop worksheet. As shown, the Hazop worksheet can include columns for 'existing safeguards' and 'how indicated' to link into modern concepts of safety management.

The amount of documentation produced during a hazard identification study should not be used as an indicator of its effectiveness; indeed, too much effort in this area may well detract from the real purpose.

A good indicator of the quality of the study is to consider the technical content and originality of the questions raised by the team. For example, the questions raised in the review of a furnace by a team which does not include a furnace expert may be of much narrower scope and more superficial than those of a team which includes such an expert.

2.5   COMPUTER AIDS AND FUTURE DEVELOPMENTS

There are many computer software programs and aids available[7,13]. Computers have simplified the documentation and reporting of hazard studies and and allow the convenient development of a database of actions for subsequent follow-up and implementation.

There are several interesting projects underway to incorporate fundamental or past knowledge in an automated or semi-automated manner in the hazard identification process, using expert system or database technologies. Currently these are not able to replace the experience and creativity of the team, but it is likely that they may play an increasing supplementary role. Examples of projects to develop knowledge-based computer aids are STARS and STOPHAZ which are co-ordinated by the CEC Joint Research Centre at Ispra, Italy and ICI respectively[14] and funded by the Euopean Union. Other significant examples include projects in the USA and Korea.

2.6   CONCLUSIONS

Hazard identification procedures are probably the best-developed element of risk analysis, having been widely used for almost three decades in the process industries. Thus, the future will probably not see much fundamental development of the methods, but rather a wider application and more emphasis on 'tools to facilitate their application'.

Industrial motives for the development of knowledge-based computer aids include:
- reproducibility;
- removal of Hazop from the critical path of capital projects;
- reduced dependence on the availability of experts.

Academic motives include:
- exploration of the capabilities of qualitative physics and qualitative modelling;
- knowledge elicitation, representation and archiving;
- industrial application of state-of-the-art artificial intelligence techniques.

Most research aims at full emulation by computer of conventional Hazop as a long-term objective. There is however an expectation that the most likely outcome will be the use of computers to improve conventional Hazop as a team support, rather than its complete substitution.

REFERENCES IN CHAPTER 2

1. Turney, R.D. and Roff, M.F., 1995, Improving safety, health and environmental protection on existing plants — process hazards review in *Proceedings of the 8th Intl Loss Prevention Symposium, Antwerp, 6–9 June* (Elsevier Science) 93–105.

HAZOP WORKSHEET			
PID nr 1, part 1	Line/equipment: kerosene unloading From: kerosene truck To: TK102		
Guide word	Deviation	Possible causes	Consequences
No	no product, no kerosene	truck (nearly) empty	• vacuum in the truck, truck collapses, spill → fire etc • pump overheats and gets damaged, gets on fire
No	no flow of kerosene	drain valve open, leakage of joint	• environmental problem, fire in the sewer, danger to personnel
No	no flow of kerosene	rupture of flexible hose	• spillage → fire
No	no flow of kerosene	valve closed	• pump overheats, pump gets on fire
No	no flow of kerosene	freezing of line, line blocked	• line bursts → spill, fire etc
No	no flow of kerosene	vacuum in truck	• collapse of the truck
No	no flow of kerosene	kink in the hose	
More	more kerosene	more kerosene for transfer than room available in TK 102, larger truck than normal 20 m^3	• TK 102 will overflow, cause fire, environmental problem
More	more flow rate of kerosene	truck pressurized	• overpressure of tank, TK 102

Figure 2.3 Example of a Hazop worksheet.

HAZARD IDENTIFICATION PROCEDURES

		Page: 1
Design intention: Transfer content of truck, kerosene waste to TK 102	Liquid waste plant	DATE: 30.9.93

How indicated?	Existing safeguards	Action required	Nr
	supervisor check (Safe Operations Procedure, SOP)	• complete and extend the SOP transportation procedure • sampling procedure needs to be completed • check design of trucks	1 2 3
visible	operator	• consider installing toewalls, sump • SOP, operating preocedure, check for leaks	4
visible	operator	• operator watch unloading	1
Pl in line		• identify system to prevent that valves are closed during pumping, SOP	1
		• operating procedure foe cleaning of lines with N_2	1
		• check truck design, vacuum breaker	3
	metallic hoses	• check SOP • maintenance of hoses	1 5
LAH + SOP	SOP check volume	• consider LSH cut out for pump P. 102, local LAH, sound alarm • consider design TK 102 with weak seam	6 7
		• check PSV capacity • check design of truck	8 3

2. Kletz, T.A., 1984, *Cheaper, Safer Plants or Wealth and Safety at Work. Notes on Inherently Safer and Simpler Plants* (IChemE).
3. The DOW Chemical Company, 1994, *Dow's Fire and Explosion Index Hazard Classification Guide and Chemical Exposure Guide*, AICHE Technical Manual, 2 vols, LC 80-29237 (AIChE).
4. *The Mond Index*, 1985, 2nd Ed (ICI).
5. *Check-lists Processing Plants*, V 11 E 2nd Ed. 1989, and P172-2 (DGA, The Netherlands).
6. *A Guide to Hazard and Operability Studies*, 1977 (Chemical Industries Association, London).
7. *Guidelines for Hazard Evaluation Procedures* 2nd ed, 1992 (Center for Chemical Process Safety, AIChE).
8. Solomon, C.H., The Exxon chemicals method for identifying potential hazards, *Loss Prevention Bulletin,* 52, August 1983.
9. Kirwan, B. and Ainsworth, L.K., 1992, A Guide to Task Analysis (Taylor & Francis).
10. Kletz, T., 1993, *Lessons from Disaster* (IChemE).
11. Lees, F.P., 1980, *Loss Prevention in the Process Industries* (Butterworths).
12. Turney, R.D., 1989, Procedures for the control of safety, health and environment hazards in the design of chemical plant, *International Conference on Safety and Loss Prevention in the Chemical and Oil Processing Industries, Singapore.*
13. *CEP Software Directory.* A supplement to *Chemical Engineering Progress*, annual (AIChE).
14. Rushton, A.G., 1993, *Knowledge Based Hazops Parts 1 and 2* (European Process Safety Centre).

3. CONSEQUENCE ANALYSIS

3.1 INTRODUCTION TO CONSEQUENCE ANALYSIS

Consequence analysis is the part of risk analysis which considers the physical effects of hazards and the damage caused by them. It is carried out in order to form an opinion on potentially serious hazardous outcomes of accidents and their possible consequences.

The purpose of consequence analysis is therefore to act as a tool in the process of decision making in a safety study which involves the following steps:
- description of the system to be investigated;
- identification of undesirable events (discussed in Chapter 2);
- determination of the magnitude of the resulting physical effects;
- determination of the damage;
- estimation of the probability of occurrence of calculated damage (discussed in Chapter 4);
- assessment of risk against criteria (discussed in Chapter 5).

The results of consequence analysis are useful to a variety of people and institutions.
- The chemical and process industries get information about all known and predictable effects that are of importance when something goes wrong in the plant, and also get information on how to deal with possible (credible) catastrophic events;
- The designers get information on how to prevent or minimize the consequence of accidents;
- The workers in the plant and the people living in the vicinity of the plant get an understanding of their personal situation. It will probably be important for them to know that safe technology surrounds them;
- The legislative authorities.

Consequence analysis should be performed by professional technologists and physicists who are experienced in the actual problems of the technical system. Figure 3.1 (page 30) shows the logic chain of consequence analysis in the process of decision making is given in . This figure is generally applicable for most process releases, but important exceptions exist (for example, less common routes to boiling liquid expanding vapour explosion (BLEVE) or chemical transformations upon release).

The first step in the chain is a description of the technical system to be investigated. In order to identify the undesirable events, a scenario of possible incidents needs to be constructed. The next step is to carry out model calculations which take damage level criteria into account. Then, after discussion, conclusions can be drawn.

Feedback from model calculations to the scenario is included, since the linking of the outputs from the scenario to the inputs of models may cause difficulties. There is also another feedback: from damage criteria to model calculations,

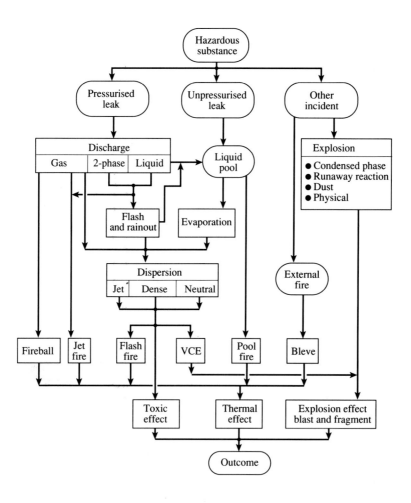

Figure 3.1 Logic diagram for consequence analysis.

in case these criteria should be influenced by possible threshold values set by the legislative authorities.

3.2 EFFECT MODELS

Physical effects can be calculated by means of effect models in which the vulnerability of the environment is not taken into account. The characteristic features of these models are described in the rest of this chapter.

3.2.1 DISCHARGE

In the past ten years much of the work related to discharge modelling has been undertaken for pressure vessel blowdown prediction.

Blowdown calculations account for how the source pressure decreases with time as the inventory in a system is depleted. Blowdown may occur as a result of a punctured tank, a break in the associated piping, or the lifting of a pressure relief valve. The last of these three can occur as a result of a runaway reaction or external fire attack. When volatile liquids are present, the internal flashing evaporation process cools the system, and the pressure decrease tends to follow the vapour pressure curve for the material in the system. Non-reactive systems may be kept initially above their vapour pressure curve by the use of non-condensable padding gas. Reactive systems may generate non-condensible gases.

When the puncture or pipe break occurs well below the liquid level in a vessel, the discharge is single-phase, sub-cooled liquid, or two-phase, saturated or flashing flow. When the opening occurs above the liquid level, the discharge is initially single-phase vapour/gas. Shortly thereafter, though, the discharge can become two-phase, with both non-condensible gas and condensible vapour in one phase and liquid as the other. As the liquid level in the vessel falls, a different disengagement regime may dominate and flow again becomes all vapour/gas.

Liquids stored above their saturation temperature at one atmosphere tend to form bubbles upon depressurization. With the formation of bubbles, the volume change forces the liquid level to swell. The extent of the swell is determined by the bubble rise velocity and the equilibrium that is established between bubble rise and liquid disengagement. This phenomenon, termed partial vapour disengagement, has been broken down into four flow regimes:
- bubbly;
- homogeneous or foamy;
- churn-turbulent;
- entrained-droplet.

The bubbly regime occurs with high viscosity systems in which bubbles move relative slowly (in laminar flow). Homogeneous flow is more vigorous, typical of turbulent flow, yet with poor disengagement. With better disengagement, bubbles agglomerate into slugs in the — even more — vigorous churn-turbulent regime. With large vessel 'freeboard' volume, vapour is the continuous phase, and liquid is carried over as droplets.

Blowdown models
Analytical solutions are available for blowdown for single-phase systems. For sub-cooled liquids these solutions treat nearly all vessel geometries of interest. For gases which do not reach condensation temperatures, the blowdown solutions can be shown on a single plot. Both blowdown and discharge models are summarized by Woodward[1].

For two-phase blowdown the situation is too complex for analytical solutions and requires a computer-based numerical model.

Discharge rate considerations and model types
For many simple discharge calculations, the classic Bernoulli equation may suffice for pure liquids and the sonic discharge equation for pure gases.

A general case discharge occurs from a hole smaller than the pipe diameter in a pipe emitting two phases including non-condensible gases, condensible vapours, and liquids. Special cases occur when:
- the hole size equals the pipe diameter;
- the pipe length is zero (orifice or nozzle flow);
- non-condensible gases are not present;
- only a single phase, liquid or gas/vapour, is present.

In each case, for compressible flow (gas/vapour and/or two-phase flow) for most practical values of source pressure, the mass discharge rate occurs at choked-flow, independent of downstream conditions. Choked-flow occurs when sonic velocity is reached in the orifice and thus reductions in downstream pressure will not increase the flow. The mass rate continues to increase, however, with increasing upstream pressure. The choked flow point is very distinct when the discharge is through a smooth diverging/converging nozzle. The choked-flow point is less distinct for sharp-edged orifice or pipe break conditions. Nevertheless, theory developed for constant entropy discharge is used here, for lack of alternatives.

For compressible flow, the cross-sectional area of the discharge decreases at first to form a minimum at the vena contract and then enlarges. The flow velocity accelerates to reach sonic velocity at the vena contract in the case of single-phase gas/vapour flow. With two-phase flow, the velocity at the choke point may not be precisely the equilibrium two-phase sonic velocity, though, because liquid vaporization requires nucleation and a time for bubble growth to reach vapour/liquid equilibrium. Furthermore, as the vapour expands in the pressure gradient, the vapour velocity exceeds the liquid velocity until drag forces equalize them. The slip ratio of vapour to liquid velocities reduces therefore to one, termed homogeneous flow. When vapour/liquid equilibrium is reached, the flow regime is termed homogeneous equilibrium.

Experiments have shown that a fairly short length of pipe — about 10 cm — is required for the discharge flow regime to reach or closely approximate to homogeneous equilibrium. Thus, discharge from moderate-to-long pipes can be adequately modelled for two-phase flow by using a homogeneous equilibrium model (HEM). Discharge from pipes shorter than 10 cm or from orifices or nozzles is better modelled, however, with either a non-equilibrium model (NEM) or a slip-equilibrium model (SEM). For critically important calculations, a more general type of model has been developed which solves five or six ordinary differential equations accounting for transfer rates of mass, energy and momentum between the liquid and vapour phases.

A key parameter in estimating discharge rates is the coefficient of discharge, C_d. For incompressible flow, C_d has a generally accepted value of 0.61. For compressible flow, C_d varies with vapour quality (mass fraction) from 0.61 to about 0.95 for pure vapour.

The HEM applied to orifice, nozzle or short pipe (non-equilibrium) discharges provides a lower bounding approximation for the discharge work.

A particularly simple NEM was developed by Fauske. This adjusts the flash fraction, x, for non-ideality as a quadratic function of the equilibrium flash fraction, x_e. Several forms of SEMs have been developed, based on correlations for the slip velocity ratio, K, as compiled by Chisholm[2].

Experimental discharge rate data are available for a wide range of operating pressure for horizontal pipes, orifices, and nozzles. Data were developed most extensively for water. Less extensive, but adequate, data are available for other materials (ammonia, chlorofluorocarbons, ethylene, chlorine, liquid metals).

Finally, as typifies a well-developed field, the nomenclature used with discharge and blowdown modelling is becoming fairly well standardized. A useful definition has been developed for a dimensionless mass flux:

$$G^* = G / (\rho_o P_o)^{1/2}$$

where ρ_o is the possibly two-phase material density at the source conditions and P_o is the initial pressure. Values of G^* are the order of unity over a very large range of source conditions. Data expressed in this form are independent of the system of units.

An important point to note when assessing the potential conservatism of a discharge model is the application. A model that is 'conservative' for the relief valve application will tend to underestimate the flow rate through an orifice for any given fluid condition. Design calculations based on a specified depressuring time will thus tend to be on the safe side as the discharge will be greater than calculated. Such a model is 'non-conservative' when applied to loss of containment, as the parameter of interest is the maximum extent of the hazard envelope and underestimating the true discharge rate will underestimate the hazard envelope.

3.2.2   AEROSOLS

Factors influencing the dispersion and downwind effect distance following loss of containment are the mass fraction of liquid flash in the release, entrainment as an aerosol, and rain-out. Superheated liquids flash upon release, and in flashing tend to break up the liquid into fine droplets which settle out slowly. Even subcooled liquids break up into droplets when ejected under pressure. Small droplets do not rain out, but rather remain suspended in the cloud and, in the near field, either grow by condensation and water absorption (hydrogen fluoride and ammonia) or, more typically, evaporate as the vapour cloud disperses downwind. The liquid droplets increase cloud density dramatically compared with a pure vapour, and even as they evaporate they cool the cloud, thereby increasing cloud density. Measurements have shown cooling of up to 38°C below the atmospheric boiling point. It is crucial to account for the density increasing effects of aerosols in obtaining a realistic estimate of consequence distances, especially for materials such as hydrogen fluoride and ammonia which would otherwise simply lift off as a buoyant vapour.

Herman[3] reviewed aerosol formation and dispersion for the American Petroleum Institute (API), and Johnson[4] summarized an experimental programme conducted for the CCPS and US Environmental Protection Agency (EPA) on aerosol rain-out rates. Rain-out fractions were measured for several materials (CFC–11, chlorine, methylamine, cyclohexane, and water). Aerosol evaporation and condensation is quite well understood, although sometimes quite nonideal thermodynamics treatment is required (particularly with ammonia, hydrogen fluoride, and organic acids which form hydrogen-bonded oligimers in the vapour phase and react with water). Likewise, droplet drag and falling velocities are quite well studied. But very few measurements have been made on the initial drop sizes formed during an accidental release. The uncertainty in being able to calculate the initial drop size makes the entire treatment of aerosols quite uncertain, as was discovered by the API experiments with hydrogen fluoride releases in Nevada in 1987.

3.2.3 EVAPORATION

Evaporation of liquids on land
Evaporation of liquid pools on land may be classified in two ways: first by the presence or absence of a bund containing the liquid, and second by whether or not the liquid is boiling. In calculating the rate of evaporation it is necessary to know the surface area of the pool. In the case of a bund, this can be taken as up to the area of the bund. In the absence of a bund, the rate of spreading of the liquid needs to be modelled.

For boiling liquids, the calculation of evaporation rate per surface area on a non-penetrable substrate can generally be based on the heat transfer of cooling down a semi-infinite medium. If the substance is penetrable this influence can be taken into account by incorporating a correction factor in the models that are generally used for the non-penetrable case. The same is true for the effect of the freezing of any water present in the capillaries.

For non-boiling liquids the most important parameter is the mass transfer coefficient. This quantity is calculated on the basis of an idealized wind profile. In practice, however, the wind profile is always influenced by such obstacles as walls, buildings and trees[5].

For the rate of spreading, most models assume that the liquid spreads in a circle, and are valid only for flat, non-penetrable substrates. Sometimes an estimate of the minimum layer thickness is made depending on the roughness of the substrate. In practice, a liquid is unlikely to spread in a circle, due to inclination of the substrate or irregularities in its surface.

A weakness of all these models is a lack of validation against experiments. The evaporation rate models are only based on small-scale experiments carried out mainly for liquified natural gas (LNG) and LPG. Large-scale tests on chlorine are presently underway in the US.

Evaporation of liquids on water
For liquefied gases on open water the evaporation depends on the temperature difference between the boiling liquid and the water. For cryogenic fluids this temperature difference is large and heat transfer is characterized by film boiling. For most boiling liquids on open water, however, heat transfer is caused by transition or metastable boiling.

In the literature information is only available about metastable boiling of liquefied gases on solid surfaces, so these are the only results that can be applied to evaporation on water. The results obtained through this transformation have been compared to small-scale experiments with LPG, and the agreement seems to be satisfactory, although the influence of disturbances on the water surface is not taken into account. Recently, progress has been made by developing a spreading-boiling model for instantaneous spills of LPG on water; the model is based on a moving boundary heat transfer model.

When liquefied gases are spilt onto a confined water surface, evaporation is accompanied by the formation of an ice layer. The calculated results of evaporation of boiling liquids on such a layer have been verified by small-scale experiments in Dewar vessels with LPG and liquified ethane.

In summary, substantial experience and knowledge is available on evaporation of liquids on water, but is limited in the large scale.

3.2.4 DISPERSION

Dispersion models apply over the whole regime from the release point through to the final downwind concentration of interest. In principle, for a typical pressurized fluid, this covers the initial momentum-dominated phase, followed by a dense gas phase where gravitational effects dominate, until finally a passive neutral buoyancy regime applies where the gravitational effects are negligible compared to atmospheric turbulence[6-8]. As previously noted, even process fluids with molecular weights lighter than air can behave as dense gases due to temperature and aerosol effects. Some dispersion models incorporate all three regimes, and others only the dense gas and neutral gas regimes; a separate turbulent-free jet or spray release model is necessary to initialize these. Releases that are neutral or positively buoyant would bypass the dense gas model.

Dispersion models are often categorized as continuous or puff (instantaneous) depending on the type of release being modelled. The transition is not always clear-cut, as instantaneous spills are often characterized by a large initial puff followed by a continuous release generated from an evaporating pool.

Finally, most dispersion models today are primarily flat terrain models, with variable surface roughness to accommodate different local situations, but are not designed to stimulate the effect of obstacles or hills[9].

Gaussian dispersion models

Many models are based on the Gaussian dispersion model originally developed by Pasquill and Sutton for describing the far-field, long time-scale behaviour of steady, continuous emissions[10]. Further so-called puff models exist for instantaneous and 'short-duration' emissions.

These models are appropriate for neutrally buoyant plumes. Additional methods have been developed by Briggs and others to account for molecular or thermal buoyancy effects, such as those from power station chimneys[10]. Gaussian models only apply to heavy gas dispersion if the wind speed is sufficiently high, and at sufficient distance downstream for the details of the source behaviour to be unimportant. However, many toxic or flammable releases are heavier than air, either because they are at low temperature due to sudden depressurization, or because they contain liquid droplets. Gaussian models which neglect buoyancy effects are therefore of limited value.

Consequently, various models of heavy gas dispersion have been developed for analysing behaviour near the release location.

Heavy gas dispersion models

Heavy gas releases are usually classified according to two ideal types: continuous sources and instantaneous sources. The first type of source arises when there is a steady release of gas for a sufficiently long period of time for transient effects to disappear. Time variations in the plume are then entirely due to stochastic variations — for example, in the wind direction. From the second type of source, a 'puff' of heavy gas is formed and moves downwind, growing and diluting with downstream distance.

Methods for estimating the effects of a heavy gas release range from correlations based on trials to simplified box models and computational fluid dynamics (CFD) models to laboratory simulations. In most cases the modelling approach (box or CFD) is used and only these are considered here.

Box models generally assume a simple shape for the cloud, within which the concentration is uniform. Mass transfer is assumed to occur by entrainment across the density interface of the cloud. Entrainment velocities are assumed to depend on turbulence levels and density differences, and on cloud speed. They are determined from laboratory experiments or by comparing model predictions to data from field experiments. Box models thus solve simple equations for a small number of variables which are functions of downwind distance only. They can be applied to either instantaneous or continuous releases. Other models — for example, SLAB — are similar to box models but allow treatment of transient releases by making the variables functions of time as well as downwind distance[11,12].

Three-dimensional models numerically integrate suitably simplified equations of mass, momentum and energy conservation. In principle, such models can incorporate terrain effects and complex geometries, and describe wake effects around structures. Simplification of the equations involves time-averaging to obtain mean values for the concentration, velocity, pressure and temperature fields. As in all turbulence models, the averaged equations need to be closed by making assumptions concerning the fluctuating correlations, such as the use of eddy diffusivities or the k-epsilon model (named after its two parameter formulation).

As a heavy gas cloud dilutes it reaches a point where it is no longer heavy — that is, its density becomes close to that of the surrounding air. If the cloud is still flammable or toxic at this stage, it is important that the transition from negative to neutral buoyancy is adequately modelled. Some models assume a specific, and sudden, transition while others account for the transition naturally and continuously.

Until recently, the following assumptions were generally made:
- The dispersing cloud moves over flat terrain or water.
- The ground or water has constant roughness and constant thermal properties.
- There are no obstructions such as dykes, tanks or buildings to the wind or moving cloud.
- The dispersing gas undergoes no chemical or physical reaction during dispersion.
- Local concentration fluctuations are not predicted.

Currently, the heavy-gas box-models are the most common and are widely used in risk analysis. CFD techniques are more difficult to apply and are less well validated, mainly due to the large number of case-specific assumptions required.

Recent developments
Since 1987, the Commission of the European Communities (CEC) has been funding work by European scientists on heavy gas dispersion. The work[8] has included application of simple box models to the calculation of concentrations downwind of obstacles, using a virtual source technique. These simple models have also been extended to take into account concentration fluctuations. Also, intermediate two-dimensional models based on the 'shallow water' (or 'shallow layer') approximation have been developed. Such models present a potentially valuable means of avoiding the prohibitive costs of full three-dimensional simulations, while allowing modelling of complex cloud shapes not treated by the simpler box models. In this way the effect of obstacles and of sloping terrain can now be modelled. Additionally, exothermic reaction with atmospheric water is treated. Interest is growing in the important case of a transient source — that is, one that is intermediate between a continuous and instantaneous source. The particular case of hydrogen fluoride (oligomerization, highly exothermic reaction with atmospheric water) has recently been addressed in dispersion models.

Validation and evaluation of models
Many field trials have been performed in Europe and the US. European Union (EU) work has included the Thorney Island trials of 1980, and more recently

two major series of field tests in parallel with wind-tunnel modelling. The first series included over 100 releases of propane. The second series involves field-scale releases of ammonia and continues. This work has brought many advances in experimental technique, and resulted in the establishment of data sets. A CEC-funded project is underway to store these data in an accessible format for the validation of models[13,14].

Validation is not the only factor involved in evaluating the quality and reliability of dispersion modelling. Other factors include a scientific assessment of the model and a verification of correct coding. In response to widespread concern about the quality of models used in risk assessments, the CEC has set up the model evaluation group (MEG), whose purpose is to raise awareness of the problem and to improve the culture in which models are developed and used. The MEG includes a specialized working group on heavy gas dispersion. This group has conducted a classification of models and data, produced an evaluation protocol and conducted an informal evaluation exercise. It is expected that the MEG will have an important role in quality assurance of models in this area in the future.

Ideally, from the viewpoint of consistency and ease of use, consequence models linking the discharge regime, aerosols, rain-out and re-evaporation directly to the dispersion model are preferred since they reduce interfacing problems and simplify validation.

3.2.5 THERMAL RADIATION

The treatment of burning materials in risk analysis can be separated into two main parts: the source term (thermal heat flux, smoke generation and other harmful loads) and effects on nearby vulnerable items (mainly people, structures, and the potential for escalation). The potential for ignition is also an important issue in many risk assessments.

There are many distinct types of fire situation which require different types of analysis to estimate consequences accurately. A basic categorization would include the following types:

- flash fire or cloud fire — the delayed ignition of a flammable gas or vapour cloud which, in the absence of significant confinement or obstruction, results in a low velocity flame front with minimal overpressure effects and primarily local impacts. If the fuel source is still available, a flash fire event is likely to continue as a jet or pool fire. If obstructions or confinement are present, flame front acceleration is possible and a vapour cloud explosion may result (see page 43).
- jet fire — the ignition of high velocity gas or liquid flammable releases, which are characterized by high momentum and good combustion conditions in the open. A variant on jet fires are diffusive fires which are characterized by

lower exit velocities and dominated by thermal buoyancy effects rather than momentum. A liquid or two-phase release will generate a burning spray as well.
- pool fire — a fire of flammable or combustible liquids burning on a flat horizontal surface (solid or liquid), often characterized by poor combustion and large generation of smoke.
- fireball/BLEVE — a fireball is the surface-burning of a cloud of unmixed flammable gas usually following the rapid release from pressurized containment (see page 42). Some overpressure may arise from BLEVEs due to the rapid physical expansion of the initially liquid fuel.
- ventilation-controlled fires — fires within buildings or compartments where the intensity of the fire is determined by the rate of ventilation, and hence access to oxygen, rather than by the availability of fuel.

Modelling of thermal effects

In principle the heat radiation of burning pools, flash fires or flare stacks and BLEVEs can be calculated on the basis of models containing the flame shape and size, radiation intensity (surface heat flux of the flame), the geometric view factor and atmospheric transmission. When calculating the view factor an important problem may be to determine the shape and time evolution of the flame.

TNO[15] and CCPS[16] provide useful surveys of current fire modelling techniques in use in the process industry, and SINTEF[17], Norway, extends this to cover special issues associated with offshore installations. In general terms, the modelling of pool fires and BLEVEs has been relatively well established for some time, whereas substantial effort has been devoted more recently to characterizing jet fires and to smoke generation and movement.

Currently, most fire modelling is characterized by extensive use of correlations for flame dimensions and estimates for surface emitted flux (the thermal radiation intensity of the flame surface). It is unusual at present to employ fundamental radiation modelling, such as the Stefan–Boltzmann law, due to the lack of knowledge of actual flame temperatures in accidental events and the high sensitivity of that formulation which involves absolute temperatures to the fourth power.

Pool fires

Pool fires are typically modelled starting with an estimate of pool size. Then the radiant heat received at a target location can be calculated using a burning rate to generate the heat source, a flame height correlation, a wind tilt correlation, a surface emitted flux (often including a factor to account for smoke obscuration), a geometric shape factor and an estimate for atmospheric transmissivity (mainly a function of atmospheric humidity).

The effect of the geometrical shape of the flame and the place and orientation of the exposed object are incorporated in the view factor. The view factor represents that fraction of the radiated energy that falls on a particular target. It is determined by geometry, primarily the shape and orientation of the flame surface and target and the distance separating them. Pool fire models are steady state simulations. This is reasonable for longer-term exposures, but large diameter pool fires are often pulsing in nature due to the effect of wind, air replenishment and smoke dynamics.

The intensity of radiation of a flame is dependent on the temperature, the type of fuel and the diameter of the pool. Experimental data on thermal intensity is available for a range of materials up to 10 m diameter, for LNG up to 35 m, and for kerosine up to 50 m. The relatively limited experimental data show a relatively large spread in the degree of the radiation intensity. Some experimental results of LNG and LPG pool fires on water were obtained during the China Lake and Maplin tests and experimental results of LNG, LPG and kerosene pool fires on land[15–17]. The surface emitted flux is strongly affected by the degree of soot formation. Smoky flames may generate only 20–50% of the radiation associated with bright non-smoky flames. Typical flame surface thermal fluxes from large-scale pool fires of typical hydrocarbon liquids are usually 140 kWm^{-2} or less, often around 50 kWm^{-2} for smoky fuels, reducing in the limit to 20 kWm-2 for totally smoke-obscured fires. LNG and LPG burn with hotter, less smoky flames and these can generate flame surface fluxes over 200 kWm^{-2} fluxes.

Special attention is required for pool fires which are floating on water as these are unconstrained by a bund and the water will modify the burning rate, especially for liquefied gases (notably LNG and LPG). Sea surface fires can be serious for marine tankers and for offshore platforms as the metal hull or support structure can fail under thermal impact. SINTEF[18] has carried out a review of these factors.

Jet fires

Jet fires have received substantial attention in the past ten years, mainly to address fire risks on offshore platforms due to closely-spaced process equipment and where weight restrictions greatly increase the cost of passive protection. Radiation from jet fires is normally characterized by the visible flame dimensions. Surface emitted thermal radiation from large (10–20 kg s^{-1}) jet fires of offshore natural gas have been measured at 200 kWm^{-2}.

Models devised for safe flare radiation separations in *API 521*[19] and by Wertenbach, while suitable for these purposes, were not sufficiently accurate for design purposes at close range. This is because small differences in flame dimensions and surface-emitted flux can result in quite different local radiation intensities. The Shell Thornton Research Centre (TRC) in Chester, UK, has carried out a number of full-scale jet fire radiation experiments using fuels typically found on offshore installations and the TRC model of Chamberlain[20] is well regarded.

BLEVEs (Boiling Liquid Expanding Vapour Explosions)
BLEVEs normally result from the catastrophic failure of a pressure vessel containing liquid flammable gases under pressure. Often the failure is the result of a fire beneath the vessel. A strongly radiating, rising fireball is created. While the blast damage is usually mild, thermal radiation damage can be substantial. The available data for fuel fireballs with respect to size and time agree reasonably, but the theories are not detailed enough and have some questionable points.

Most BLEVE models are based on correlations for quasi-steady diameter, plume rise, duration and surface emitted flux. These are combined with geometric view factors and atmospheric transmissivity to estimate thermal impacts at various target locations. Unlike pool fires, BLEVEs are of short duration, often 15–45 seconds, depending on BLEVE mass. The largest uncertainties are associated with the surface-emitted flux.

Recently, greater attention has been given to estimating BLEVE hazards from pressure vessels used in transport. This is because of the greater proximity of such hazards to the general public (see 'Transport Risks' page 108).

Special cases
Most fire models have been developed to address the open field situation. A major concern in many process situations, especially for offshore installations, is the confined space hydrocarbon fire. In such cases it is necessary to determine if the fire will be fuel-controlled or ventilation-controlled. In the latter case, standard jet fire or pool fire models may be inadequate, and allowance must be made for fire beyond the module vents.

Smoke hazards are now receiving increased attention, rather than purely thermal radiation impacts. This is primarily due to the Piper Alpha disaster, where the majority of fatalities were due to smoke, rather than due to direct thermal radiation. Smoke possesses several hazards: toxic components (mainly carbon monoxide), oxygen depletion, high temperature breathing effects and vision obscuration. Carbon monoxide concentrations of 3–4% are typical in poorly ventilated fires. Smoke soot production is variable but values of 7–10% are typical for heavier hydrocarbons in poorly ventilated situations.

The distribution of smoke from a fire is complex and several attempts have been made to simulate movement in various geometries using CFD codes such as FLOW–3D, KAMELION and PHOENICS. This technology is not yet suitable for routine wide-scale application due to limited validation, costs and the expertise required. There have been several models developed to characterize smoke ingress into vulnerable areas (for example, control rooms, temporary safe refuges): AIRQUAL (Shell), SMILE (AEA Technology), FAST (US Center for Fire Research).

3.2.6 VAPOUR CLOUD EXPLOSIONS

Substantial work has been undertaken in the explosion area, both on the source term and on the resulting damage. Good reviews are presented by the IChemE[21], Roberts and Pritchard[22] and CCPS.

A vapour cloud explosion may be simply defined as an explosion occurring outdoors, producing a damaging overpressure. It begins with the unplanned release of a large quantity of flammable vaporizing liquid or gas from a storage tank, process vessel or pipeline. Generally, several features need to be present for a vapour cloud explosion with damaging overpressures to occur. These are:
- the released material must be flammable;
- a cloud of sufficient size must form prior to ignition (dispersion phase). Should ignition occur instantly, a large jet fire or fireball may occur — itself causing extensive localized heat radiation damage — but significant blast pressures causing widespread damage are unlikely to occur.
- a sufficient amount of the cloud must be within the flammable range of the material to cause extensive overpressure. A vapour cloud will generally have three regions — a rich region near the point of release, a lean region at the edge of the cloud and a region in between that is within the flammable range.
- the blast effects produced by vapour cloud explosions are determined by the speed of flame propagation. This implies that the mode of flame propagation will be a deflagration. Under extraordinary conditions a detonation might occur. In the absence of turbulence, the vapour cloud will simply burn and the event is described as a large flash fire. Therefore an additional condition is necessary for vapour cloud explosions with pressure development: turbulence. Research testing has shown turbulence significantly enhances the combustion rate in deflagrations.

Turbulence in a vapour cloud explosion may arise in two ways, namely by the release of flammable material or by the presence of multiple obstacles — for instance, pipework — within a process plant. Both mechanisms may cause very high flame speeds and, as a result, strong blast pressures. The generation of these high combustion rates is limited to the release area or to congested areas

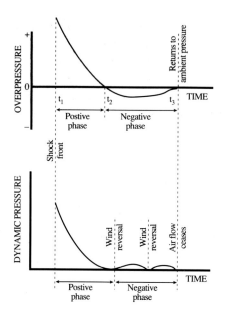

Figure 3.2 Variation of overpressure and dynamic pressure with time, at a fixed location, for an ideal blast wave[21].

respectively. As soon as the flame enters an area without turbulence or without congestion, the combustion rate drops as well as the pressure production.

Historically, the phenomenon was referred to as an 'unconfined vapour cloud explosion' (UVCE). It is, however, more accurate to call this type of explosion a 'vapour cloud explosion' (VCE). A graph showing the overpressure and dynamic overpressure associated with a VCE, as given in the IChemE's *Explosions in the Process Industries* monograph[21] is given in Figure 3.2.

In the extreme, the turbulence can cause a sufficiently energetic mixture to transfer from deflagration to detonation, with much greater damage potential. It should, however, be emphasized that for a detonation to propagate, the flammable part of the cloud must be very homogeneously mixed. This transition is not well understood.

Explosion prediction models
Basically there are two groups of models describing explosion blast. The first group of models quantifies the source strength as an equivalent quantity of high explosives, generally trinitrotoluene (TNT), in order to be able to apply well-known TNT-blast characteristics. Historically, this group of models is widely used and well accepted.

During the last decade, the processes involved in a vapour cloud explosion have been the subject of intense international research after it was found that some incidents could not be represented by the methods available[22]. This research effort resulted in methods which take into account the different behaviour of a vapour cloud explosion compared to a high explosive detonation. This second group of models will be referred to as fuel-air blast charge models and are being applied, although some lack of data hinders a general acceptance.

Methods based on TNT-blast
For many years, the military has been interested in the destructive potential of high explosives. Therefore, it is understandable to relate the explosive patterns observed in many major vapour cloud explosion incidents to equivalent TNT-charge weights. Basically, the use of TNT-equivalence methods for blast predictive purposes is very simple. The available combustion energy in a vapour cloud is converted into an equivalent charge weight of TNT. The factor used to relate hydrocarbon combustion energy to TNT is referred to as the equivalence factor, yield factor, efficiency or efficiency factor.

From the equivalent TNT-charge weight, the blast characteristics in terms of the peak side-on overpressure of the blast wave at any distance from the charge can be estimated and the corresponding damage pattern determined.

Practical values for the TNT-equivalence of vapour cloud explosions are much lower than the theoretical upper limit. For most of the major vapour cloud explosion incidents, TNT-equivalencies range from 1% to 10%, based on the heat of combustion of the full quantity of fuel released[23]. Apparently, only a relatively small part of the total available combustion energy is actually involved in generating an explosion overpressure.

Brasie and Simpson[24] have published a well-known TNT-equivalence model. They base their recommendation for the TNT-equivalence of vapour clouds on the damage observed in three chemical plant explosion incidents. Analysing the damage in these incidents, they deduced a TNT-equivalence which is highly dependent on the distance to the explosion centre.

Although it was recognized that, occasionally, much higher values have been observed in vapour cloud explosion incidents, surveys show that most major vapour cloud explosions involving hydrocarbons have in practice developed between 1% and 3% available energy present in the cloud. A TNT-equivalence of 3% is suggested in general. The TNT-equivalence should be increased for more reactive gases: for above average reactivity (for instance, propylene oxide) 6% should be used, and for very reactive gases (ethylene oxide) 10% is suggested.

One of the complicating factors in the use of a TNT-blast model for vapour cloud explosion blast modelling is the distance-dependency of the TNT-equivalence observed in incidents. Properly speaking, TNT-blast characteristics do not correspond to gas explosion blast. For the representation of far-field blast effects a much heavier TNT-charge is required than for near-field effects.

In all TNT-equivalence models the basic assumption of a proportional relationship between the amount of fuel available in the cloud and the TNT-charge weight expressing the cloud's explosive potential is most questionable. This is reflected by the wide range of equivalencies found in the analysis of a large number of vapour cloud explosion incidents involving fuels whose heats of combustion are of the same order of magnitude as hydrocarbons.

Nevertheless, the TNT-equivalence concept makes it possible to model the blast effects of a vapour cloud explosion in a very simple and practical way. The great attractiveness of these methods is the very direct, empirical relation between a charge weight of TNT and the attendant structural damage. Therefore, the TNT-equivalence is a useful tool if the property damage potential of vapour cloud explosion is the major concern. It has the additional benefit that different analysts are likely to produce broadly similar estimates of damage.

Methods based on fuel-air blast
In TNO's Yellow Book[15], the gas dynamics induced by a spherical expanding piston are used as a model for vapour cloud explosion blast. A piston blast model offers the possibility to introduce a variable initial strength of the blast.

This approach makes it possible to model vapour cloud explosion blast by considering the two major characteristics of gas explosion blast:
- the scale, determined by the amount of combustion energy involved;
- the initial strength, determined by the combustion in the explosion process.

Experimental research during the last decade has confirmed clearly that deflagrative combustion generates blast only in those parts of a quiescent vapour cloud which are sufficiently obstructed and/or partially confined. The remaining parts of the flammable vapour-air mixture in the cloud burn out slowly, without significant contribution to the blast. This idea is called the multi-energy concept and underlies the method of blast modelling[25]. Figure 3.3 shows what lies behind this model.

In this method a vapour cloud explosion is defined as a number of sub-explosions corresponding to the various sources of blast in the cloud. Blast effects are represented by using a blast model. Generally, blast effects from vapour cloud explosions are directional. Such an effect, however, cannot be modelled without a detailed numerical simulation. In the multi-energy method

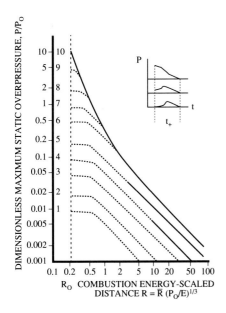

Figure 3.3   TNO multi-energy method: overpressure[25]. (Reproduced by permission of Elsevier, Amsterdam, The Netherlands.)

a simplified approach is used in which blast effects are represented in an idealized, symmetrical way with the BLAST-code.

The initial strength of the blast is indicated by a number ranging from 1 for very low to 10 for detonative strength. The initial blast strength is defined as a corresponding set of blast wave parameters at the location of the charge radius. In addition, a rough indication for the shape of the blast wave is given. Pressure waves produced by a fuel-air charge show different decay characteristics in the near and far fields.

In the multi-energy method the explosion hazard is determined above all by the environment in which the vapour disperses and the boundary conditions to the combustion process. If a release of fuel is anticipated the explosion hazard assessment can be limited to an investigation of the effects of the environment on the potential blast.

Another method that also uses the multi-energy concept as the starting point is described by Kinsella[26]. It uses the same set of blast charts as the multi-energy method. Kinsella argues that due to the assumption in the multi-energy method of a stoichiometric homogeneous cloud, there are three factors left which govern the severity of a VCE:

- degree of congestion by obstacles inside the vapour cloud;
- ignition energy;
- degree of confinement.

To estimate the initial blast strength for particular types of explosion, a review has been carried out of major accidents where near-field explosion overpressure has been estimated from structural damage and where information is available regarding the level of congestion, confinement, release size and ignition strength.

The Kinsella approach offers the possibility to differentiate explosion source strengths depending on the initial and boundary conditions. At the present time the method suffers from the limitation that the guidance provided on the choice of class number for the explosion strength is more limited than for the TNO approach. Unfortunately, no comparison has been presented between the incidents that were evaluated and the blast overpressures predicted.

Although multi-energy methods have proved that they are capable of explaining the broad range of effects observed in real explosions, they are still very dependent upon the initial assumptions made by the analyst. For this reason, at the present time the range of answers which can be obtained is very large and this limits the usefulness of the method. In order to improve guidance and increase consistency, a project known as GAME (Guidance for the Application of the Multi-Energy method) has been undertaken at TNO Prins Maurits Laboratory.

Methods based on numerical simulations
These methods can model the necessary fluid dynamic and combustion processes for explosion simulation. The 3-D codes are capable of simulating the basic mechanism of a gas explosion, which is the interaction between combustion, expansion and turbulence. This has resulted in a number of equations and sub-models that have to be solved for each grid point and each step in time.

These equations and sub-models are:
- conservation equations for mass, momentum and energy to model the gas dynamics;
- equations for the conversion of reactants into combustion products;
- a sub-model to determine the combustion rate based on turbulent mixing of reactants and products;
- a sub-model for turbulence consisting of equations for turbulent kinetic energy and its dissipation rate;
- a sub-model for turbulent premixed combustion;
- a sub-model for a laminar burning speed controlled start;
- a sub-model for sub-grid obstacle representation.

These codes can be applied in all situations where the effects of a vapour cloud explosion have to be considered. Applications have been found in studies of overpressure to be expected in a range of typical modules on off-shore platforms.

At the time of publication, the codes available and under development are EXSIM[27], FLACS[28] and REAGAS (TNO). It is essential that these codes are validated against experiment, including the CEC projects MERGE and EMERGE.

3.3 VULNERABILITY MODELS

3.3.1 INTRODUCTION

In earlier chapters it has been shown that the effect caused by an undesirable event can be calculated by means of effect models. The damage caused to people, facilities or the environment can be assessed by means of so-called vulnerability models. It is most conveniently discussed in terms of explosion damage, fire damage and toxic damage. Table 3.1 (page 50) gives the place of the respective vulnerability models in the assessment of damage. (See also Reference 29).

In some cases other media may be involved in the cause and effect chain. For instance, a spillage of toxic material could, under unfavourable circumstances, contaminate land, affect underground aquifers and — potentially — pollute drinking water supplies.

Most of the vulnerability models in use at the moment are restricted to the effect on people or structures. However, increasing interest in environmental aspects is leading to the development of additional vulnerability models and relationships.

The following sections (3.3.2 to 3.3.4) contain brief descriptions of the characteristic features of the vulnerability models used for calculating explosion damage, fire damage and toxic injury. Any comparison between calculated theoretical results and the limited experimental data is beyond the scope of this book.

3.3.2 EXPLOSION DAMAGE

Structural damage
An explosion invariably causes a blast wave which, if strong enough, may result in structural damage. Decisive for the amount of damage are:
- the properties of blast wave;
- the interaction between blast wave and structure;
- the properties of the structure.

Table 3.1
Scheme of damage assessment

Damage causing event	Cause of injury or damage	Vulnerable resource	Type of injury or damage	Acute (immediate) or chronic (long-term)
Toxicity	Toxic gases and vapours, dusts or liquid	People, fauna	Death Non-lethal injury Irritation	Acute or chronic
		Flora	Death Restricted growth	Acute or chronic
Explosion	Direct blast	People, fauna	Death Non-lethal injury (eardrum rupture)	Acute
		Flora	Destruction (trees)	Acute
		Structure	Structural damage Glass breakage	
	Flying fragments, impact, structural damage from above	People, fauna	Death Non-lethal injury	Acute
Pool and flash fires, fireballs (BLEVE)	Thermal radiation	People, fauna	Death Non-lethal burns	Acute
		Flora	Destruction Ignition	
		Structures	Ignition	

One of the effects of an explosion is a sudden pressure rise. This pressure rise moves, in the form of a wave, from the centre of the explosion. The shape of this wave depends on the magnitude of the explosion and on the distance from the centre of the explosion. In the case of a detonation, the pressure rise is essentially instantaneous, and thus without a rise time; it is called a shock wave. Shock waves and pressure waves are jointly called blast.

In a shock or pressure wave, the maximum pressure rise, following a determined path, decreases to zero within a given time, followed by a time period with negative pressure. The maximum value of this negative pressure does not normally play any important role, since it is relatively gradual and its maximum value is generally low when compared to the peak overpressure. For these reasons, this negative phase is often disregarded (see Figure 3.2, page 44).

When the wave front meets a structure, reflection takes place and a refraction wave is formed. After this, the blast envelops and passes over the structure. Considering reflection, two limiting cases can be distinguished. If the direction of the blast wave is perpendicular to the surface, the pressure is maximum (normally reflected loading), but when the direction is parallel to the surface of the structure, no pressure increase occurs (side-on loading).

Structures which are loaded, one way or another, deform. The manner in which these deformations take place, as well as their values, depends not only on the loading but also on the properties of the structure. These properties are determined, in turn, by the materials used for the different structural components, as well as the manner in which the components are put together. Structures can also vibrate, in which case their own periods of vibration play an important role.

The types of loads considered here, coming from an explosion, can be very large in comparison with the loads which the structure is calculated to withstand under normal conditions. With occupied buildings, it is important that complete failure does not occur, although when dealing with such exceptional load conditions it is possible to accept a certain degree of damage in the form of permanent deformation. A property of the structure which plays a role in this respect is its ductility.

Models to predict structural damage
No universal model which takes account of all important factors for the prediction of damage to structures is available. The methods used at present are:
- Empirical pressure and pressure-impulse criteria — one method to determine the strength of structures is an empirical approach. Practical 'rules of thumb' can be found in literature which provide values of overpressure corresponding to a given degree of damage. A drawback of such rules comes from the fact that the types of structures and damage levels described in them are given in very general terms. Furthermore, only overpressures are mentioned, but it can be shown that impulse is also important in dynamic analyses.

A closer examination of these empirical rules shows that many of them come from Glasstone[30] and apply to nuclear explosions. A blast load resulting from a nuclear explosion is very different to that from a vapour cloud explosion.

Therefore, values obtained from these references must be used with care. Wells[31] gives a summary of blast damage criteria for a range of process equipment (Figure 3.4).

The most reliable data appears to be that given by Jarret. A great number of well-described explosions and records of damage to housing due to bombing are presented in his papers[32]. Thanks to this information it is possible to establish a relationship between the quantity of explosives and the distance to the centre of the explosion at which a given degree of damage has taken place. Since the parameters of a shock wave — peak overpressure and positive phase duration or impulse — are, at a certain distance, determined by this quantity of explosives, it is then possible to establish a connection between these parameters and a given degree of damage. A study performed by Baker[32], resulted in the so-called pressure impulse diagram. The advantage of this approach is that pressure and impulse determine the level of damage without a TNT-equivalence concept.

- Modelling of structures — In order to be able to determine analytically the response of a structure, a schematic representation of the structure is necessary.

A dynamic calculation provides data about the manner in which a structure reacts to a dynamic load. Such a calculation can require a great deal of calculation work. Quite often the only values of interest are the maximum forces and the maximum displacements of a structure. In order to determine them in a simple fashion, the so-called 'quasi-static' calculation procedure can be used. In this procedure, the maximum value of the dynamic loading is multiplied by a given factor, whereby it becomes possible to calculate the loading using static design methods. The factor by which the maximum dynamic load must be multiplied is called the dynamic load factor (DLF). The DLF is determined with the help of the dynamic response calculation for a single-degree-of freedom system, and is primarily dependent on the duration of the dynamic loading. In Biggs *et al*[33], values of the DLF are calculated.

For elasto-plastic behaviour of structures, the ductility also plays a role in the determination of the maximum displacement. Reference 34 gives diagrams, for a number of load schemes, from which the maximum response can be determined. These diagrams can be used in a number of different ways.

If data about the load are known (peak overpressure and positive phase duration), the required combinations of DLF and ductility which are necessary to prevent failure of a structural system can be found.

If, on the other hand, data about the structure are given (ductility, static strength), the type of dynamic load that the structure can withstand can be found.

Finally, if all data are known, an impression about the safety (or lack of it) of a given structure can be obtained.

CONSEQUENCE ANALYSIS

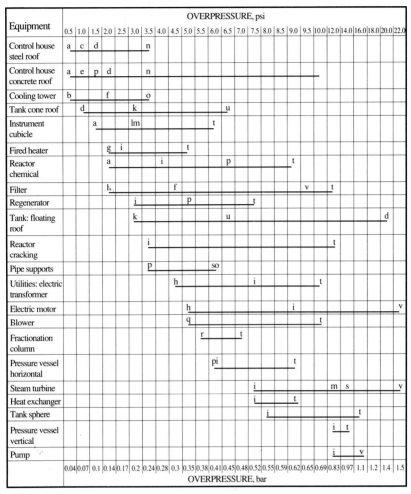

Figure 3.4 Typical damage caused by overpressure effects of an explosion. Reproduced with permission of the US Department of the Interior, Office of Oil and Gas.

a Windows and gauges break
b Louvres fall at 0.3–0.5 psi
c Switchgear damaged by roof collapse
d Roof collapses
e Instruments damaged
f Inner parts damaged
g Brick cracks
h Debris-missile damage occurs
i Unit moves and pipes break
j Bracing fails
k Unit uplifts (half-filled)

l Power lines severed
m Controls damaged
n Block walls fail
o Frame collapses
p Frame deforms
q Case damaged
r Frame cracks
s Piping breaks
t Unit overturns or is destroyed
u Unit uplifts (0.9 filled)
v Unit moves on foundation

53

- Computational modelling — Blast waves generated by vapour cloud explosions decay in the surrounding atmosphere. The flow field within a blast wave has a complicated structure, particularly when it is influenced by the presence of rigid boundaries. Often, numerical simulation is the only way to assess the blast loading on surrounding objects with some accuracy. Examples include BLAST-3D (TNO), FLACS (Christian Michelsen Research), and CHAOS (British Gas).

Damage to humans

It is customary to identify three categories of explosion damage to humans depending on the causative mechanism of damage.

The first category is primary damage caused by direct blast effects, the primary cause of lethality being lung haemorrhage. As the external pressure on the chest wall is larger than the internal pressure, the chest moves inward. This causes the injury. Most of the criteria for lung injury found in the literature are extracted from Bowen[35]. Reference 36 gives an overview of calculated chances of survival.

The main non-lethal injury resulting from direct blast effects is eardrum rupture. Eardrums are damaged in response to overpressure alone, since the characteristic period of the ear vibration is small compared with the duration of a blast wave. Reference 36 gives relationships between the overpressure and the probability of eardrum rupture.

The second category is secondary damage caused by missiles and fragments. In the determination of the effects of fragments on the human body, a distinction is made between cutting and non-cutting fragments.

The injuries which may result from cutting fragments are lacerations and punctures. Cutting fragments are often light (10 grams or less in mass), flying glass being one of the most common examples.

Non-cutting fragments, which are mostly large objects, cause internal injuries in humans. It is not possible to predict theoretically the damage caused by non-penetrating objects on the human body with any accuracy. In Reference 36 a figure was composed based on the scarce and incomplete data available on the subject.

People inside collapsing buildings are subjected to the impact of very heavy structural parts. From pictures taken after earthquakes or bomb attacks, it can be seen that the vertical structural members usually fail, leaving a stack of floors on top of each other. A number of people survive such catastrophes, probably due to the co-incidental creation of arches which give protection.

Where the TNT-equivalent model is used relationships are available linking the proportion of occupants likely to survive to the calculated overpressure. These relationships can be applied to buildings of 'normal construction', like houses and offices. Care should, however, be exercised in the case of industrial buildings. For instance, at Flixborough the control building had a solid concrete floor above the control room. No-one survived the collapse of this building.

The third category of explosion damage to humans is tertiary damage by transmission and subsequent collision with an obstacle. The air particles in the blast wave have a velocity which is, in general, in the same direction as the propagation of the blast wave. People can be carried along by this explosion wind and thrown against obstacles. This sweeping away mainly occurs if the human body is in an upright position.

Based on the pressure and impulse of the incident blast wave, the maximal velocity can be calculated for a human body during transportation by the explosion wind. Reference 36 gives the lethality criteria for whole body impact as a function of overpressure and impulse, based on available data in literature.

3.3.3. FIRE DAMAGE

Damage resulting from heat radiation covers damage to structures and injury to humans. For all types of damage two parameters have been found to be significant — the level of thermal radiation and its duration.

Fire damage to structures

Fire damage can range from paint flaking off to the ignition and burning of the object. In the case of a non-combustible material, the temperature can increase to the point at which the material loses strength and stiffness. If such a material is used in load-bearing constructions, it is possible that the construction will collapse at a given heat load.

For damage to structures one key concern is whether ignition occurs. Since surface treatment, geometrical position and other factors play important roles in the determination of ignition, in Eisenberg[29] the problem has been simplified by only considering the ignition of wood. Here, reference is made to experimental results in which the intensity of radiation capable of igniting wood spontaneously both with and without a pilot flame has been determined.

The Green Book[36] indicates critical radiation intensity values for wood, plastics, glass and uncovered steel. This critical radiation intensity value is defined as the maximum value at which no ignition occurs, regardless of the length of exposure time.

Modified 'view factor' methods are used to translate the emitted radiation from a flame to the received radiation at a surface.

RISK ASSESSMENT IN THE PROCESS INDUSTRIES

Fire damage to humans
Injury caused to humans by fires is mostly characterized as first, second and third degree burns and lethality. The severity of the injury can be calculated from the given heat radiation, starting from a known exposure duration and radiation intensity (see Figure 3.5). One approach frequently used is to find the LD_{50} (lethal dose) radiation level appropriate to the time for which the individual will be exposed.

Much consequence analysis is based on simple thermal flux criteria which are determined from an assumed exposure time. These times have been derived from animal experiments or from piloted or unpiloted ignition of combustible materials.

Empirical relations are available in which a type of injury is expressed in probit (probability unit) functions which are described in more detail on page 60. However, most probit functions are based on thermal pulses from nuclear weapons. As the wavelength of thermal pulses caused by heat radiation from nuclear weapons is within the ultraviolet (UV) part of the spectrum, these functions cannot be applied to estimate thermal injury caused by hydrocarbon pool

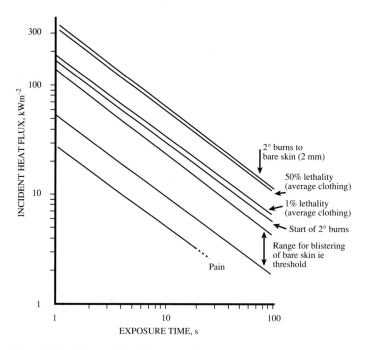

Figure 3.5 Injury and fatality levels for thermal radiation. (Reproduced with permission of AEA Technology. Taken from I. Hymes, report *R275*, 1983).

fires etc. Therefore, the probit functions in Reference 36 are modified so that they can be applied to heat radiation within the infrared (IR) part of the spectrum (pool fires, flash fires, etc.).

Clothing can have a protective influence for humans, until the moment it ignites. In order to estimate this influence, a relation has to be known between age, percentage of body area burnt and mortality rate. Reference 36 discusses such a relation; it appears that, assuming about 20% of the body area remains unprotected for an average population, the lethality is 14% of the lethality for unprotected bodies.

When probits are used, it is important to limit the exposure duration, otherwise low thermal fluxes (for example, below solar radiation levels) can be predicted to lead to thermal injury.

Another aspect that influences the heat radiation injury significantly is the effect of escaping to a safe location. This can be incorporated by changing the exposure time into the effective exposure time. Reference 36 derives a relation for the effective exposure time based on information in the literature.

These models make no mention of smoke inhalation or suffocation, although these effects are often injurious or even fatal to humans trapped by fire.

3.3.4 TOXIC INJURY

As noted in Table 3.1 (page 50) toxic injury may arise from the effect of gases, vapours, dusts or liquids with the effects ranging from irritation through to fatality.

The modelling of toxic injury constitutes a highly complex problem since it depends on a number of varying factors, such as:
- the intrinsic properties of the toxic materials under consideration (for example, acute toxicity versus chronic toxicity);
- the mode of action of such materials (for example, reversible versus irreversible effects, local versus systemic effects);
- the mode of entry into the human system (inhalation, skin-absorption, ingestion, etc);
- the dose received by an exposed individual as a function of exposure time and concentration of the toxic agent;
- the variablility of individual response to toxic exposure;
- the limitations in the applicability of experimental results obtained from animals and micro-organisms to humans and the scarcity of valid epidemiological data;
- the uncertainty about the existence of no-effect levels for carcinogens and mutagens.

Unfortunately such detailed information is rarely available. Relationships between exposure conditions and the effects are generally based on animal experiments and need to be modified to relate to humans[37]. In many cases even the animal experimental data may be based on very small samples.

Local and systemic effects

As already noted, chemicals may act on the body in two ways to give local or systemic effects.

Locally-acting substances cause damage at the point where the substances contact the body — for instance, the lungs in the case of a gas or vapour. When interpreting animal data, account must be taken of differences in respiration volume, lung surface, breathing pathway, etc.

Systemic-acting substances cause injury to the body via the blood distribution within the body. In some cases a specific organ may be affected.

Simple approaches to toxic effects

One of the simplest approaches was that first used by Dicken[38] in ICI to cover the toxicity of chlorine in 1974. Data collected from the literature covering adult

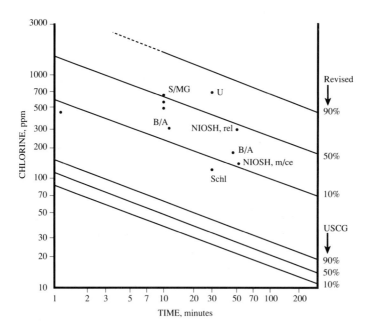

Figure 3.6 Toxic fatality probability diagram for chlorine. B/A = Bilron and Aharonson (mice); U = Underhill (dogs); USCG = US Coast Guard; S/MG = Silver and McGrath (mice); Schl. = Schlagbauer and Henschler (mice). (Reproduced with permission of Akzo Engineering bv, Arnhem, The Netherlands.)

casualties to industrial gassing accidents were collated and presented as graphs of concentration (ppm) versus duration (mins) for broad categories of annoyance/smell, harmful and dangerous. Although very simple, the approach has the benefit that a wide range of effects can be considered and is still used today in a modified form.

Where animal experiments have been carried out, the results may be expressed as the LC_{50} — the concentration required to kill 50% of the species for the nominated exposure period. Such data needs specialist skills in acquiring and translating to the effect on humans.

Many consequence analyses are carried out for emergency response planning. For these a toxic concentration indicative of onset of injury over a short period is appropriate. Historically, the best known such toxic concentration is the IDLH (immediately dangerous to life and health) published by the US National Institute for Occupational Safety and Health (NIOSH) for a wide variety of chemicals assuming a 30 minute exposure duration. More recently, the US National Academy of Sciences has produced carefully reviewed three level ERPG (emergency response planning guideline) concentrations for a smaller number of chemicals. These have an exposure duration of 1 hour.

Toxic dose

Concentration/time relationships (see Figure 3.6) can often be approximated by straight lines on a log–log plot. The relationship is then conveniently represented by an expression of the form:

$$c^n t = \text{constant}$$

c = toxic gas concentration (usually given in parts per million or mg m^{-3})
t = time of exposure (usually given in minutes)

The slope of the line, n, is referred to as the 'toxic index'. The value of n is usually between 1 and 3 and depends on the manner in which the gas produces a toxic effect. For instance, it is generally accepted that n is greater than 1.0 for chlorine because this gas attacks by two mechanisms: immediate respiratory spasm from intense irritation and delayed effects from oxidative and other destructive effects on lung tissue. The first effect is a function of concentration and the second of dosage; hence as concentration increases the acute effect becomes more important and progressively dominates the other.

For risk analysis purposes one convenient approach is to use the concentration/time equation to define a 'toxic dose'. In some cases industry working groups have been established to pool experience and determine the most appropriate toxicity relationship to use. Such a group established by the European Chlorine Industry (Eurochlor) under chairmanship of Dr R. Papp studied chlorine toxicity and recommends the relationship prepared by Vis van Heemst[39] (Figure 3.6).

Probit equations

The concept of a toxic dose below which a given effect will not occur is a simplifying assumption. Strictly it would be more realistic to say that the probability of the hazardous effect will become less as the dose decreases. Account can be taken of this by using a probit method.

In this approach, for toxic gases, a probit Y is defined by an expression of the form:

$$Y = A + B \ln c^n t$$

Where sufficient information is available A and B are determined for a defined effect (usually death). The probit Y is a random variable with a Gaussian distribution; the values of A and B are chosen such that Y has a mean value of 5 and a variance of 1. Y provides a measure of the proportion of the population who would be expected to be adversely affected by a toxic release causing a toxic dose $c^n t$. Values of the probit Y are calculated from this equation and transformed to the probability (0 to 1.0) of the hazardous effect occurring by a theoretical relationship that can be conveniently shown graphically. A general description of the theory and application of the probit method is given in Lees[10].

For example, the chlorine toxicity data shown in Figure 3.6 (page 58) can be represented by the following probit where c is the concentration in mg ppm and t is the time in minutes:

$$Y = 0.5 \ln c^{2.75} t - 4.0$$

In theory the probit method is applicable to all toxic materials but difficulties arise in practice. The derivation of probit functions for human lethality is severely restricted by the fortunate shortage of appropriate toxicity data on which to base a reliable judgement for the values of the parameters in the probit expression. There is a wide diversity in probits published in the literature.

Probit equations for a number of chemicals have been developed by the Toxicity Working Party of the IChemE Major Hazards Assessment Panel[40] and have been published by the IChemE. In addition, a detailed study undertaken in Holland has resulted in the production of probit equations for 22 chemicals[36].

Long-term effects

Repeated, long-term exposure to harmful chemicals may give rise to health effects. Legislation in a number of European countries places an obligation on the employer to identify such substances and ensure that, in the methods used, the risks are reduced to an acceptable level.

Limits for long-term exposure may be obtained from the US threshold limit values (TLVs)[41], German MAK values[42] and UK occupational exposure limits (OELs)[43]. The European Union is currently developing common indicative limit values (ILVs).

Effects on the environment
As noted earlier, data on the effects of chemicals on the environment are at present rather limited. There is further discussion in section 6.5 on page 119.

3.3.5 CONCLUSIONS

The effect of the many variables involved in assessing toxic effects and the many conservative assumptions built into some studies mean that even the most detailed representations, such as probits, need to be treated with care. While data for a small number of chemicals have been extensively peer reviewed (for example, NAS ERPG values and the IChemE Major Hazard Assessment Panel review), this is not so for the majority of chemicals. Furthermore, newer data, collected using better protocols, may require previously established toxic criteria to be updated. The advice of an expert toxicologist is therefore essential in assessing the validity of data for a particular study, especially where less common chemicals are involved.

REFERENCES IN CHAPTER 3
1. Woodward, J.L., 1993, Discharge rates through holes in process vessels and piping in *Prevention and Control of Accidental Releases of Hazardous Gases* (Van Nostrand Reinhold).
2. Chisholm, D., 1983, *Two-Phase Flow in Pipelines and Heat Exchangers* (George Godwin, London). Published in association with IChemE.
3. Herman, M.N, 1987, Aerosol formation, and subsequent transformation, and dispersion during accidental releases of chemicals, *API Publication No. 4456* (American Petroleum Institute).
4. Johnson, D.W., 1991, Prediction of aerosol formation from the release of pressurized superheated liquids to the atmosphere, *CCPS Conference, New Orleans, May 20–24* (AIChE), 1–44.
5. Center for Chemical Process Safety, 1989, *Guidelines for Chemical Process Quantitative Risk Analysis* (CCPS, AIChE).
6. Britter, R.E., 1989, Atmospheric dispersion of dense gases, *Annual Review of Fluid Mechanics*, 21, 317–344.

7. McQuaid, J., 1989, in *Methods for Assessing and Reducing Injury from Chemical Accidents*, Bordeau, P. and Green, G. (eds) (Wiley).
8. Wicks, P.J. and Cole, S.T., 1995, European research in accidental release phenomena, *International Conference and Workshop on Modelling and Mitigating the Consequences of Accidental Releases of Hazardous Materials, New Orleans, 26-29 September*.
9. Webber, D.M., Jones, S.J., Martin, D., Tickle, G.A. and Wren, T., 1994, Complex features in dense gas dispersion modelling volume 1, *report no. AEA/CS/FLADIS/1994* (AEA Technology Consultancy Services).
10. Lees, F.P., 1980, *Loss Prevention in the Chemical Industries* (Butterworths), 207.
11. Kukkonen, J, Kulmala, M., Nikmo, J., Vesala, T., Webber, D.M. and Wren, T., 1994, The homogenous equilibrium approximation in models of aerosol cloud dispersion, *Atmospheric Environment*, 28, 2763–2776.
12. Duijm, N.J., 1995, Dispersion of dense gas and flashing releases, *International Conference and Workshop on Modelling and Mitigating the Consequences of Accidental Releases of Hazardous Materials, New Orleans, 26-29 September*.
13. Wicks, P.J. and Cole, S.T. (eds), 1994, Seminar on the evaluation of models of heavy gas dispersion, Mol, Belgium, 25 November, *EUR report 16146 EN* (Commission of the European Communities).
14. Hanna, S.R., Chang, J.C., Strimaitis, D.G., 1993, Hazardous gas model evaluation with field observations, *Atmospheric Environment*, 27A, 2265–2285.
15. TNO, 1988, *Methods for the Calculation of Physical Effects Resulting from Releases of Hazardous Materials. Yellow Book*, 2nd ed (TNO).
16. Center for Chemical Process Safety, 1994, *Guidelines for Evaluating the Characteristics of Vapour Cloud Explosions, Flash Fires and BLEVEs* (CCPS, AIChE).
17. SINTEF, 1992, *Handbook for Fire Calculations and Fire Risk Assessment in the Process Industry* (with Scandpower, SINTEF, Trondheim).
18. Stensass, J.P., 1992, Fire on sea surface — state of the art and the need for future research, *Report STF25 A92035* (SINTEF, Trondheim).
19. *API Recommended Practice 521*, 1990 (American Petroleum Institute).
20. Cowley, L.T. and Johnson, A.D., 1992, *Oil and Gas Fires Characteristics and Impact,OTI 92-596* (Steel Construction Institute, London).
21. *Explosions in the Process Industries. Major Hazards Monograph*, 2nd edition, 1994 (IChemE).
22. Pritchard, D.K. and Roberts, A.G., 1993, *Blast Effects from Vapour Cloud Explosions. A Decade of Progress* Safety Science 16, 527–548.
23. Gugan, K., *Unconfined Vapour Cloud Explosions* (IChemE).
24. Brasie, W.C. and Simpson, D.W., 1968, Guidelines for estimating damage explosion, *63rd National AIChE Meeting, St. Louis* (AIChE).
25. van den Berg, A.C., 1985, The multi-energy method — a framework for vapour cloud explosion blast prediction, *Journal of Hazardous Materials*, 12.
26. Kinsella, K.G., 1993, A rapid assessment methodology for the prediction of vapour cloud explosion overpressure, *Proceedings of the International Conference and Exhibition on Health, Safety and Loss Prevention in the Oil, Chemical and Process*

Industries, Singapore 15–19 February (Butterworth-Heinemann) 200–211.

27. Hjertager, B.H., Solberg, T. and Nymoen, K.O., 1992, *Journal of Loss Prevention in the Process Industries*, 5.
28. Bjerketvedt, D. and Bakke, J.R., 1992, *Gas Explosion Handbook Volume 1, CMR-92-F25036* (Chr. Michelsen Research)
29. Eisenberg, N.A. et al, 1975, *Vulnerability Model, A Simulation System for Assessing Damage Resulting from Marine Spills* (US Department of Transportation).
30. Glasstone, S., 1975, *The Effects of Nuclear Weapons* (US Atomic Energy Commission).
31. Wells, G.L., 1980, *Safety on Process Plant Design* (George Godwin, London), 3. Published in association with IChemE.
32. Baker, W.E. et al, 1979, *A Short Course on Explosion Hazard Evaluation* (Southwest Research Institute San Antonio).
33. Biggs, J.M. et al, 1959, *Structural Design for Dynamic Loads* (McGraw-Hill).
34. van den Berg, A.C. and Lannoy, A., 1993, Methods for vapour cloud explosion blast modelling, *Journal of Hazardous Materials*, 34, 151–171.
35. Bowen, I.G. et al, 1966, *Biophysical Mechanisms and Scaling Procedures Applicable in Assessing Responses of the Thorax Energized by Airblast Overpressures or by Non-Penetrating Missiles, DASA 1857* (Lovace Foundation for Medical Education and Research).
36. *Methods for Determination of Possible Damage to People and Objects Resulting from Releases of Hazardous Materials (Green Book)*, 1989, (Committee for the Prevention of Disasters, Directorate-General of Labour of the Ministry of Social Affairs, The Hague).
37. Fairhurst, S. and Turner, R.M., 1993, Toxicological assessments in relation to major hazards, *Journal of Hazardous Materials*, 33, 215–227.
38. Dicken, A.N.A., 1974, *The Quantitative Assessment of Chlorine Emission Hazards* (Electrochemical Society, San Francisco).
39. Vis Van Heemst, M. and ten Berge, W.F. 1983, Validity and accuracy of a commonly used toxicity assesment model in risk analysis, *4th International Symposium on Loss Prevention and Safety Promotion in the Process Industries, Harrogate, 12–16 September* (IChemE), I1–I12.
40. *Ammonia Toxicity Mongraph*, 1988, *Chlorine Toxicity Monograph*, 1989, *Phosgene Toxicity Monograph*, 1993 (IChemE).
41. *Threshold Limit Valves for Substances in Workroom Air* (American Conference of Governmental Industrial Hygienists).
42. MAK values (Deutsche Forschungsgemeinschaft
43. HSE, annual, *Occupational Exposure Limits EH 40* (HSE Books).

4. QUANTIFICATION OF EVENT PROBABILITIES AND RISK

This chapter discusses the techniques used to estimate event probabilities. The strengths and weaknesses of the analytical techniques and the data requirements for quantification are examined. Accuracy and uncertainty in these estimates are considered and the implications for use of the techniques are discussed, together with the potential benefits and limitations.

The combination of these event probabilities with the results of consequence analysis to produce estimates of the overall risk from an activity is outlined. There is a brief discussion of important risk terminology and a commentary on presentation of risk analysis results, with particular reference to open publication. The likely uncertainty in estimates of risk from process plants is reviewed.

4.1 EVENT PROBABILITY ESTIMATION

4.1.1 EVENT DEFINITION

Careful definition of the events to be quantified is an important stage in an analysis. It is particularly vital in a full analysis where the probabilities and consequences of the various possible events are to be combined to produce an overall quantitative risk estimate.

The definition of events should also reflect the purpose of the risk analysis, and how the results are going to be used. A risk analysis aimed at providing decision support for selection of operational procedures, or planning for emergency preparedness, requires different focus and degree of details than a study carried out to demonstrate the feasibility of the early plans of a concept.

In the case where loss of containment is the major concern, the breakdown of event probability against size of release is a common practice to allow coupling of probability and consequences. This is not usually treated as a continuous function since the possible releases will tend to be characteristic of the particular plant — that is, the dimensions of equipment and the conditions such as pressure and temperature under normal and fault conditions. Similar releases are usually grouped together in order to achieve the coupling of events with the appropriate consequences in the overall risk analysis.

A parallel development of QRA methodology in recent years is based on using a continuous function to represent the release size distribution and then to apply probabilistic simulation techniques to develop resulting consequences as a statistical distribution. The methodology is fairly complex, but reduces some of the pessimism often present in discrete estimates as well as offering a clearer reflection of site-specific details and practices.

4.1.2 THE PHILOSOPHY UNDERLYING EVENT PROBABILITY ESTIMATION

There are two basic approaches to event probability estimation. The first is direct use of statistical data on failure of plants or whole systems. This is sometimes called the 'historical approach'. The second is to break down the event into its contributory factors and causes, using analytical/simulation techniques.

An advantage associated with the use of historical event data is that, where the accumulated experience is relevant and statistically meaningful, the assessment will not omit any of the significant routes leading to the event. The data already encompasses all common relevant contributory aspects including the reliability of equipment, human factors, operational methods, quality of construction, inspection, maintenance, operation, environment, etc. However, outdated modes which may not be relevant to a specific case under study are also included (due to improvements to equipment, operational practices or management regimes specific to the plant), resulting in an over-estimate — usually referred to as a 'conservative' estimate — of the chance of the event. A disadvantage of the historical approach is that the historical record is more likely to be dominated by older plant. Improvements in the standards of design, construction, testing and management of a facility are unlikely to be fully reflected. Conversely better knowledge of material properties and operating experience leads to less conservatism in design specification or more extreme operating conditions.

It is common to establish total generic failure frequencies for specific process items (pipes of a given diameter, pressure vessels, for instance) and to apply to this total a cumulative hole size distribution (Figure 4.1, page 66). It might be thought that this would be discrete (associated with characteristic hole sizes), but data collection exercises have not found this.

Often the historical data is of such quality that, while undoubtedly relevant, it is not considered adequate. In this case synthesis must be used but the predicted event probability should be tested against whatever experience exists to judge whether the different approaches produce compatible predictions.

In this chapter the chance of occurrence of an event is generally referred to by the term 'probability'. This quantity may indeed be a probability

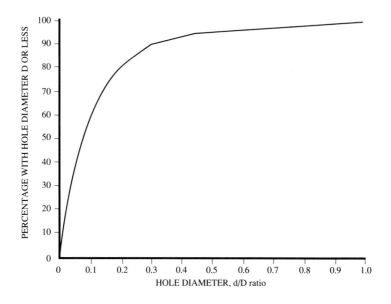

Figure 4.1   Typical hole size distribution for process piping

and therefore dimensionless. In many circumstances, however, it will be expressed as a frequency of occurrence over a specified time interval — for example, a year or a plant lifetime. The distinction between dimensionless probability values and frequency values is important where these numbers are to be used in combination.

4.1.3   TECHNIQUES FOR SYNTHESIS OF FAILURE LOGIC

There are various techniques for modelling the mechanisms — the logical combinations or sequence of events — by which an undesired event could occur. Most of these techniques are based on diagrammatic methods known as logic diagrams. The techniques are initially qualitative in nature, although they provide models for subsequent quantification if considered appropriate. Some of these techniques have already been mentioned in Chapter 2 on hazard identification and their usefulness in this role should not be underestimated. The distinction between identification techniques and failure logic synthesis techniques is therefore somewhat artificial. The Hazop and FMEA methods discussed in Chapter 2 do not provide the logical framework for setting down full event causes and effects which characterizes logic diagram approaches. However, logic diagrams must start from an event which has been identified by some method such as Hazop or FMEA.

becomes significant in the analysis, judgments can be made about whether the level of performance called for is realistic or not. The value of the risk analysis approach lies in identifying those critical situations. There is further consideration of these factors in Chapter 5.

An important trend in human factors analysis has been the recognition of the importance of organizational aspects on overall human error.

4.2.5 ACCURACY

The accuracy of a probability estimate depends not only on the uncertainty attached to the data. Two other factors are whether all of the significant contributors to the event have been identified, and whether the mathematical model applied to evaluate the event probability provides a sufficiently detailed description of the failure modes concerned.

The techniques for hazard identification and failure logic synthesis are powerful and capable of application in a wide variety of circumstances. If used rigorously they can assist the analyst in achieving a high degree of completeness. But, as with any tools, they are capable of misuse and misapplication.

The difficulty of modelling failure modes is one of many factors which tend to lead risk analysts to introduce conservative (that is, pessimistic) assumptions, resulting in an unquantified bias in a probability estimate. The presence of such elements of conservatism must be considered both when comparing estimates and, in particular, when using estimates in an absolute sense.

Data introduces possible inaccuracy in addition to uncertainty. As already mentioned, there are many assumptions underlying the use of failure data which, if not valid, could lead to a different failure rate being realized in practice and therefore an inaccurate estimate of the failure probability being made. Often directly relevant data is not available. Use of data for similar but different items is, in itself, a sweeping assumption which is commonly required. Where possible, it is important to monitor performance in practice to check whether such assumptions were correct, but the final decisions made may not have been sensitive to such data inputs.

4.2.6 USES, BENEFITS AND LIMITATIONS

Some of the benefits which derive from use of the techniques described in the section on event probability estimation, are not related to quantification. The value of the understanding of failure mechanisms and their combinations with minimal cut set analysis generated by logic diagrams should not be underestimated. In particular, they allow a thorough understanding of an activity to be built up, enabling people not familiar with that activity to bring an independent viewpoint to an entrenched procedure or operation. This assists in identification of key areas and provides an aid to communication on how systems may fail and what effect modifications might have.

Besides the uncertain contribution that quantification may make in particular cases, there are potential difficulties in its application which have been mentioned earlier in this chapter, and which must be borne in mind. The techniques themselves can appear to be extremely simple and, indeed, this is part of the usefulness of failure logic diagrams. Errors can be made, however, if an analyst is unaware of the theoretical basis underlying their construction, manipulation and evaluation, and sufficient expertise is necessary to at least be aware of such difficulties. This implies setting up a centre of such expertise in an organization — and therefore committing resources — as, at least at the outset, it is probably not feasible to train all designers to the appropriate level.

As with the hazard identification procedures discussed in Chapter 2, there is always a danger that the analysis will not be sufficiently penetrating or broad-ranging. Unless care is exercised, logic diagrams can be hardware orientated and pay insufficient regard to human factors. This is mainly a difficulty in hazard identification, but obviously it may lead to an inaccurate event frequency estimate and possibly, therefore, an incorrect conclusion.

Another factor which must be considered is cost. Cost depends on the scope and depth of a study. Detailed studies to quantify event probabilities are often time consuming and therefore expensive. Even if a study is carried out by a frequency analysis specialist, designers and operators must also be involved, so that all of the necessary knowledge and experience is applied. The initial examination of probability may be coarse and, on some occasions, this provides adequate information on which to base decisions, making costly refinements unnecessary. If this is the case, costs can be relatively modest.

4.2.7   POSSIBLE FUTURE DEVELOPMENTS

Computer codes have been developed to assist in all phases of frequency analysis: fault trees, event trees, reliability block diagrams and full probabilistic simulation. Considerable effort has been expended in the area of fault trees in the past 15 years, but it has been narrow in application and is usually associated with large complex trees.

Computer codes for automatic and semi-automatic fault tree generation are becoming widely available. These programs tend to be viewed by the risk analysis community with suspicion; the results are, of course, only as good as the logic input by the analyst. Their application would be expected to be limited to very complex systems, where development of the fault tree can be difficult. In such cases, however, correct understanding of the failure logic is usually all the more important.

The tools for synthesis and evaluation of failure logic are capable of further refinement. Computer aids are in various stages of development; the degree of progress will be fixed by demand, which is likely to be limited. New technologies may bring new problems, leading to modification of existing techniques or development of new ones.

Collaborative data projects with the co-operation of large numbers of organizations is the area where potential for progress is probably greatest. Information sharing schemes and data banks have a vital role to play, as arrangements can generally be made to overcome problems regarding commercial confidentiality. The problem of demonstrating very high reliability — that is, obtaining a statistically meaningful sample for rare events — will always remain. Data collection may well proceed independently of risk analysis, as much of the data is useful in reliability and availability studies for which there is generally a keener economic incentive.

4.3 QUANTITATIVE EXPRESSIONS OF RISK

4.3.1 TERMINOLOGY

There are many proposed definitions for the term 'risk'. In the context of assessment of the hazard posed by process plant, it is now widely agreed that the risk from an activity refers to a function of the likelihood of occurrence of possible undesired events and the magnitude of their associated consequences. Some definitions of risk refer only to likelihood, or probability. But where there is a range of possible outcomes these must also be considered. Many definitions refer to risk as the product of probability and consequences. This implies that risk can be aggregated into a single number. Although single number representations of risk may be useful for some purposes, a complete description of the nature and distribution of the risk is not provided by them where a range of outcomes is possible, particularly in terms of number of fatalities. The following discussion concentrates on such situations where the main concern is harm to people, although the method used for expressing risk depends on the objectives of a particular study.

Individual risk
The most widely-used measure is individual risk. Individual risk is the frequency at which an individual suffers a defined degree of injury. For completeness, the nature of the injury, source of the risk and identity of the individual (usually by location or job title) should be specified. Individual risk is usually

used to indicate how significant the imposition of risk is compared with the background of risk an individual is exposed to. A plot of individual risk contours on a map provides a graphic picture of the geographical risk distribution[12]. A typical example is shown in Figure 4.4. Often, however, only the highest individual risk is quoted, or the individual risk at several selected locations.

A measure of risk sometimes encountered is average individual risk. Individual risk statistics (for example, for road deaths) are averaged out over a population. In the context of risk from a plant, the average individual risk is usually calculated by dividing the expectation value by the number of people exposed to the risk, giving a single number with units of casualties per person per year. The attraction of this parameter is that it provides a direct comparison with individual risk statistics. There are often problems in defining the number of people exposed and, because the risk is not fully described, changes in the pattern of risk may be misrepresented using this parameter. (For example, increasing the number exposed can appear to reduce the risk!). Therefore, although this measure of risk may be useful for putting the overall risk into context, it should be accompanied by the more detailed risk measures from which it was derived. This conclusion applies to all single number measures of risk where multiple fatalities could occur, and is similar to that drawn in the

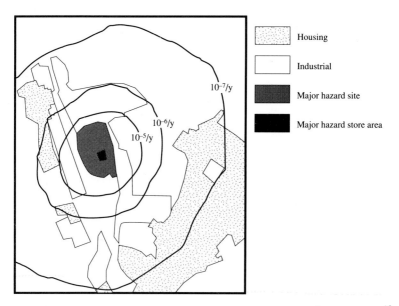

Figure 4.4   Contours of individual risk of death around a major hazard facility[13]. (Crown copyright is reproduced with the permission of the Controller of HMSO.)

report by the Royal Society Study Group on Risk Assessment[14] which recommended that, in the wider context of considering all detriments, risks and associated detriments should not be aggregated into a single index unless the contributory elements can be disentangled.

Fatal accident rate
An alternative way of expressing individual risk at work is the fatal accident rate (FAR), which allows comparison with industrial fatality statistics. The statistic is an average number of deaths per 100 million working hours, but, used as a loss prevention tool, FAR is usually applied to an individual or job function. As an individual risk measure it does not address multiple fatality accidents, although its use as a 'group risk' taking account of this aspect has occasionally been suggested. Care is then necessary in the interpretation of the significance of what is then another single number form of representation of a multiple fatality risk. The FAR concept has proved a useful tool for identifying high risks and providing a practical design target for risk reduction. Fatal accident rates can be related to individual risk mathematically.

The process industries in the UK (oil and gas production, energy production and chemical industries) have a long-term FAR of about 4 to 5. This reduces to approximately 1 if the oil and gas production sector — which includes the 167 fatalities that occurred during the Piper Alpha incident — is excluded. An FAR of 1 roughly translates to one fatal accident over 1000 working lifetimes or one per 50 yeras for a site employing 1000 people.

Societal risk
The societal risk, sometimes known as 'group risk', from an activity is the relationship between frequency and total number of people harmed. The relationship is often plotted as a cumulative frequency distribution, or 'F–N curve', giving the frequency of events exceeding any particular stated severity. Included in this concept, therefore, is the worst possible outcome — together with its frequency — and the frequency of any harmful outcome. The degree of harm considered should be specified; often this is fatality. A typical F–N societal risk plot is shown in Figure 4.5 on page 80.

Societal risk conveys the potential for disaster and all other multiple fatality events, but it does not tell us how the burden varies amongst individuals in the exposed population. Although societal risk in the form of F–N relationships can be calculated, it has proved difficult to relate this to what society finds tolerable or acceptable (see Chapter 5).

The integral of the F–N relationship, in non-cumulative form, is a single number with units of average number of people harmed per year, often

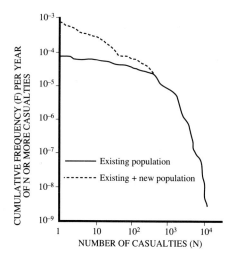

Figure 4.5 Societal risk F-N curve. Example to show effect of a new development close to the major hazard in an area which is already partly built up[13]. (Crown copyright reproduced with the permission of the Controller of HMSO.)

known as the 'expectation value'. It loses the distinction between low probability/high consequence events and high probability/low consequence events and therefore conveys less information; as a single number risk measure derived from all events, however, it is often used for cost benefit analysis.

None of these methods of expressing risk provides, on its own, a complete description where multiple fatality events are possible. The combination of societal and individual risks describes the overall risk and the distribution of the risk — concepts which are central to the taking of decisions.

Other measures of risk
In the offshore industry other measures of risk have been adopted and have proved useful in identifying and assessing effective measures to control risk. These are basically related to defined 'safety functions' which are necessary to provide for the safety of personnel. These safety functions include:
• possibility for personnel to escape to safe parts of the platform;
• integrity of safe areas (smoke and intrusion, temperature, structural integrity, and operability of safety systems);
• integrity of evacuation means.

The analysis of integrity of these functions against a range of identified source events has proved very effective in reviewing measures to prevent escalation and contain the accident in the original area, to ensure the integrity of escape routes and to specify needs for protection of the temporary safe refuge (see 'Offshore QRA' page 98).

Risk criteria focused on environmental impacts as well as material damage (damage to equipment and loss of production) are also applied to an increasing extent as a supplement to the risk criteria for personnel risk.

In addition, to broaden the perspective of the risk assessment and to focus on effective risk reducing measures, the application of these criteria often provides a wider range of arguments supporting the decision to apply safety measures.

The nature of the risk presented by an industrial activity which could conceivably cause harmful effects at considerable distances cannot be completely described in a simple way. It is therefore essential that the basis for any quoted risk is clearly stated, particularly since many of the terms used in risk analysis — including the word 'risk' itself — have less specific, everyday meanings.

When selecting the measure of risk to be used it should be remembered that the overall objective of the exercise is generally to demonstrate to the public that all reasonable measures have been taken to reduce the risks and that the remaining risks are acceptable or tolerable (see Chapter 5). Sometimes the discussion will be with experts working for the regulatory authorities but on other occasions it may be with elected representatives or members of the public. In these cases simple measures of risk which can be related to everyday experience are of greatest value.

4.3.2 COMBINATION OF EVENT PROBABILITIES AND CONSEQUENCES

The basis for a quantification of the risk from an industrial activity is a list of hazardous events, or groupings of like events, which can be considered to produce similar consequences. The frequencies of these events may be estimated by the techniques described earlier. There may be a range of possible outcomes from each event, depending on the different circumstances which may prevail — for example, wind direction, weather category and location of people. Each of the circumstances must be defined and assigned a probability. The aggregation of the frequency and consequence analysis can therefore be complex, although it is conceptually simple and all analyses follow essentially the same procedure.

Damage-causing events must be related to the initiating events — for example, the various possible outcomes arising from a release of flammable material may be modelled using an event tree. The conditional probability for factors such as wind direction towards ignition sources and chance of ignition at each source can then be used to produce a frequency for the damage-causing event from the frequency of the initiating event.

The consequences of each damage-causing event are assessed using the methods described in Chapter 3. The usual approach is to define ranges to selected casualty probabilities from a combination of effect and vulnerability models. These casualty probabilities are frequently a step function: either zero or unity. For example, a range may be quoted within which a particular level of harm is expected to be realized, as when sburn injury is based on a thermal radiation dose. Alternatively, a number of casualty probabilities may be selected and limiting ranges to each value estimated — for example, the probability of casualties occurring at various overpressures could be used in conjunction with an explosion overpressure model to produce radii to selected casualty probabilities. The selection of probabilities usually depends on the available data underlying the vulnerability model used.

Having obtained the frequency and casualty probabilities against range for each damage-causing event under consideration, the risk relationships are derived in the following manner. Taking each event in turn, the number of people present in the areas covered by each casualty probability band is multiplied by the appropriate casualty probability, producing, by summation, the total number of people predicted to be affected by each event. The overall frequency-consequence relationship can then be drawn up from the number affected and frequency for each event.

The individual risk at a location is obtained by taking the casualty probability at that location for each damage-causing event and multiplying it by the frequency of that event. The individual risk from all such events, and therefore from the activity as a whole, is obtained by summation over all the events.

The final expressions of individual and societal risk then incorporate the likelihood and severity of all the outcomes that have been considered, and allow for features specific to the plant and the particular location.

4.3.3   PRESENTATION OF RISK ASSESSMENT RESULTS

The presentation adopted depends in part on the degree of detail in the analysis and the objectives of the study. In-house studies often concentrate on the effect of design options and the degree of detail adopted is only commensurate with revealing the effect of such options. The following comments relate primarily to publications, or studies intended for third party consumption.

Societal risk can be presented in the form of a table or graph, as shown earlier in Figure 4.5 (page 80). If the 'expectation value' is used, presentation is trivial but much less information is conveyed and so the detailed F–N relationship on which it is based should also be quoted.

Presentation of individual risk is more straightforward (Figure 4.4, page 78), but because of differing definitions the results of different analyses may not be directly comparable. When the individual risk is estimated at particular locations, it is obviously necessary to specify those locations. This can be done by identifying locations on a map, but a graphic presentation is provided by risk contours. 'Averaged' individual risk is expressed as a single value with units of frequency.

It is important that the presentation is specified at the outset, as the calculations are readily incorporated into the framework in advance. Careful thought, and often much recalculation, would be required for alternative presentations once the study has been completed.

Where criteria exist, or are specified at the start of the analysis, the presentation adopted is designed to facilitate comparison with those criteria. The criteria may take the form of limiting risk values or may involve further analysis — for example, cost benefit. As these criteria cannot be regarded as absolute, it is usually best to adopt a form of presentation which gives a good overall view of the risk.

Where risk analysis results are to be published, such an overall view of the risk is essential if the results of the study are to be communicated effectively. Until such time as standardized definitions evolve, the risk terminology used in the study should always be explained.

4.3.4 ACCURACY OF RISK RESULTS

The possible accuracy of event probability estimates has already been discussed. Just as it was not possible to make generalized statements about the accuracy of such estimates, it is also not possible to make such statements about the achievable accuracy of risk estimates. The opiniions expressed about published studies can, however, be used as guidance.

In the Canvey Island Study[15] the UK HSE, commenting on the likely accuracy of the risks estimated by the investigating team, stated that the estimates of probability 'may well err on the side of pessimism by a factor of perhaps two or three, but are unlikely to err by as much as a factor of ten'. There was a similar statement concerning the consequences, implying that the probability of realization of any particular level of consequences might be overestimated by one but not two orders of magnitude. Cremer and Warner[16] considered this to be optimistic but nonetheless suggested a similar assessment of possible error: 'a factor of ten or more'.

In the Rijnmond Study[17], the level of uncertainty quoted for the risks is 'about one order of magnitude (in some cases somewhat higher, in others somewhat lower)'. An indication of the expected uncertainty in each of the case studies in the report is also given — typically 'about an order of magnitude' or 'about one-half order of magnitude'.

Other studies have failed to show such good agreement. A benchmark study carried out with the support of the European Union[18] involved eleven independent assessments of the same installation using two basic method types. The resulting risk estimates were found to differ in total by several orders of magnitude. Teams using an approach based on the Rijnmond/Canvey style of analysis (many failure cases, generic frequencies) tended towards the one order of magnitude error bands quoted before. This study emphasized the need for consistency in the methods, the models and the data used if comparable results are to be obtained. There is also an indication that techniques based on the use of failure rates from data banks lead to more consistent results.

These general levels of uncertainty cannot be assumed to apply in all cases and, in particular, reliability assessments may have much less uncertainty attached.

REFERENCES IN CHAPTER 4
1. Brand, V.P. and Matthews, R.H., 1993, Dependent failures — when it all goes wrong at once, *Journal of the British Nuclear Energy Society*, 32 (3).
2. CCPS, 1989, *Guidelines for Chemical Process Quantitative Risk Analysis*, (CCPS AIChE, New York).
3. Lees, F.P., 1980, *Loss Prevention In The Process Industries* (Butterworths).
4. FACTS database, TNO, Department of Industrial Safety, Apeldorn, The Netherlands.
5. WOAD — World Offshore Accident Database, Det Norsk Veritas, 1 Veritasvein, Hovik, Oslo, Norway.
6. MHIDAS Accident Database, MHIDAS Administrator, AEA Technology, Thomson House, Risley, Warrington WA3 6AT.
7. CEC Joint Research Centre, Institute for Systems Engineering and Informatics, *Major Accident Reporting System* (Elsevier Science).
8. *Guidelines to Process Equipment Reliability Data with Data Tables* (CCPS, AIChE, New York).
9. SRD Association, Reliability Databank, AEA Technology.
10. SRD Association, 1995, *Human Reliability Assessor's Guide, Report SRDA–R11* (AEA Technology).
11. Kirwan and Ainsworth, 1992, *A Guide to Task Analysis* (Taylor & Francis).
12. Ramsay, C.G., Sylvester-Evans, R. and English, M.A., 1982, Siting and Layout of Major Hazardous Installations, *IChemE Symposium Series No. 71* (IChemE).
13. *Risk Assessment. A Study Group Report*, 1983 and *Risk: Analysis, Perception and Management*, 1992 (The Royal Society, London).
14. Health and Safety Executive, 1978, *Canvey: An investigation* (HMSO).
15. Cremer and Warner, 1980, *An Analysis of the Canvey Report* (Oyez Intelligence Reports).

16. COVO Committee, Rijnmond Area, 1981, *Risk Analysis of Six Potentially Hazardous Industrial Objects in the Rijnmond Area: A Pilot Study* (Reidel, Dordrecht).
17. Amendola, A., Contini, S. and Ziomas, I., 1992, Uncertainties in a chemical risk assessment: result of a European benchmark exercise, *Journal of Hazardous Materials*, 29, 347–363.

5. THE APPLICATION OF RISK ASSESSMENT

Chapters 2, 3 and 4 considered how to examine a process plant for hazards, select those which could have serious consequences and calculate what those consequences might be. They also explained how to assess the probability of a hazardous event occurring and, finally, how to combine this with various possible consequence scenarios to give a picture of the risk to people and property.

This chapter summarizes the limitations associated with QRA and considers its use by an organization 'in-house' and its use in the public domain. Then it considers the way ahead.

5.1 SOME LIMITATIONS OF QRA

Before arriving at a judgement on the usefulness of risk analysis the characteristics of the assessment of probabilities and consequences have to be considered.

In many cases, only very general data are available on equipment failure, and statistical accuracy is often poor. In other cases there may be very little data available at all. This applies in particular to data on human failures. Data may have an accuracy no better than a factor of ten so that, when combined in a fault tree, they lead to incident frequencies with a wider confidence range.

It is therefore advisable to use available data from as close to the top of the fault tree as possible. For process releases which have many diverse causes, the use of generic failure data leads to more consistent estimates than failure frequencies derived from fault trees (see page 83).

There is no apparent relationship between the magnitude of the probability of a hazard and the accuracy of its estimate, as accuracy depends on the size of the sample from which it was derived. A frequent event number derived from a small sample may be more inaccurate than a rare event number derived from a much wider sample. When frequencies of catastrophic events have to be compared, use the best estimates of these frequencies to give a sound comparison.

The selection of suitable failure frequencies is a particular problem with high reliability systems, and risk assessment is usually concerned with such systems. Accurate reliability data is only likely to be available where a very large number of identical systems, many hundreds or thousands, have been in use for many years. Some systems, such as railway signals or electrical components, may meet this requirement. Process plants of a particular type are usually built in small numbers and experience rarely exceeds 20 to 30 years. Therefore generic data combined with sound expert judgement needs to be used. The methods used to quantify the effects and consequences of incidents have improved so that they can be predicted with reasonable confidence. The very large number of factors which can influence the development of an incident will, however, continue to limit overall accuracy. It will continue to be difficult to estimate whether an incident, such as the release of hydrocarbon, will have only a localized effect (that is, fire) or will escalate into a major accident. This potential for escalation tends to be addressed in greater depth for risk analysis of offshore platforms (see 'Offshore QRA' page 98).

The whole analytical exercise seeks to be objective. It must be realized, however, that as in many scientific and engineering exercises assumptions, estimates, judgements and opinions may be involved. Because of these limitations, formal quality systems are being increasingly introduced to document assumptions, employ recognized data sources and models, and thereby reduce variability. As with many techniques, they should be used by people who understand their limitations, and with caution.

5.2 APPLICATION IN THE PROCESS INDUSTRY DOMAIN

Qualitative methods for the identification of hazards have been used for many years by the process industries to ensure that their plants are adequately safe. In recent years the more intensive methods described in Chapter 2 have become more widely used. It should be remembered that these methods are used to audit a design which should already meet the many codes of practice (both national and in-house) which cover most aspects of the engineering of the plant. Whilst a number of large companies have found benefit from the use of QRA, it must be recognized that others in the process industries have not found QRA necessary. These companies, although well aware of the quantification methods described in Chapters 3 and 4, judge that the outcomes of QRA studies are not producing results on which they can rely or which contribute much to making a plant safer. They rely on identification procedures coupled with good engineering judgement, experience from actual practice and experiment, and regulations and guidelines. The availability of a large body of long-term technical experience embodied in proven codes of practice obviates the use of quantitative methods.

Consequence calculations are widely used, particularly by companies handling large quantities of flammable or toxic liquefied gases. These can be useful for determining plant siting and layout. They can also be used for planning emergency procedures. There is always a possibility, however, that too much weight is given to the largest possible consequences if a judgement of the probability of the event occurring is not taken into account (the so-called maximum credible event or worst case scenario). The proponents of quantification of risk argue that considering potential consequences alone may lead to unnecessary additions to a plant and excessive capital cost.

A company using QRA will use its own experience and judgement to define targets against which to compare the results. Such a comparison can assist it to decide — for example, on the degree of redundancy required in an instrument protection system for plant handling an exothermic reaction.

Notwithstanding the problems that still exist in the use of QRA for safety decision making, definite advantages are available if it is used prudently, particularly where new technology is involved. Benefits are to be gained in obtaining a better insight into the causes of potential plant failures. The quantification of these can help with an understanding of the relative importance of the causes, and assist with the development of improved designs. For these reasons the selective use of QRA 'in-house' is supported as one of the tools to assist with decision making on — for example, the design of a predictive system. Any organization considering a move in this direction should ensure that it has adequate expertise to handle the analytical techniques properly.

When QRA is used within a company the applications frequently involve the comparison of alternatives. In the case of a new facility the comparison may be between alternative sites. In other cases the comparison may be between alternative process routes or the way in which hazardous materials are stored. Since in these situations much of the basic data is the same for all the options being considered, the absolute accuracy of the data is not critical in making a sound decision. Even in those cases where the company may wish to compare the facility against in-house criteria of tolerability, other facilities are likely to have been studied already.

Since a detailed QRA can entail considerable work and may be costly, the scope should be tailored to the situation requiring analysis, and the depth should be no greater than that needed to make a sound decision. The extent to which risk analysis techniques are used by a company depends on its structure and management philosophy. It is as well to remember that risk analysis illuminates a problem but does not give the answer. There can, of course, be significant financial benefits when the application of risk analysis is associated with a reliability study, so that in addition to the improvement of safety the availability of the plant can be optimized.

It is also important to realize that changes made to a plant during its operating life may well invalidate the initial design studies. Therefore it is essential that such plant changes are subject at least to hazard identification.

5.3 APPLICATION IN THE PUBLIC DOMAIN

In the ten years prior to the first edition of this book, some European governments became interested in the use of risk analysis as a means of assessing whether the operations carried out on an industrial complex place local residents at too high a level of risk. Well-known studies in Europe include the two Canvey studies[1,2] in the UK, the COVO[3] and LPG Integraalstudie[4] in the Netherlands.

Many studies carried out for government authorities have concentrated on incidents with offsite effects occurring at very low frequencies. Such studies may not contribute to the prevention of smaller onsite incidents involving only a very few fatalities. Such incidents have been shown to cover the large majority fatalities related to the process industries.

Studies have enabled major hazards to be seen in perspective and in many cases they have shown that the safety of the installation or operation is acceptable to the authorities. They have also been used to suggest how risks may be reduced by improving safety measures.

Since the first edition of this book was published, there has been an increase in the use of QRA by regulatory authorities. There are, however, very considerable differences in approach between different countries. In some countries, the use of QRA has been linked to regulations for the control of major hazards such as the Seveso Directive. For example, in the Netherlands a practice has developed in which QRA is used to cover the whole range of possible outcomes of incidents to gain a balanced insight into the risks of an installation or operation. In many cases this has shown that the safety of such an installation or operation is acceptable to the authorities. In some cases it has been used to suggest how risks may be reduced, either by improving safety measures or by judicious zoning to achieve separation or to limit populations at risk[5]. In other countries — for example, the UK, a QRA is accepted as part of the safety report required under the Seveso directive, but the use of a fully-quantified risk assessment is not mandatory for onshore process facilities.

Risk analysis is increasingly used in the UK as part of the information presented at planning enquiries concerning process plant. The resultant risk figures are not being set against specific criteria, but are reviewed in relation to other risks to which the population may be subjected. The analysis is seen as part of the information required to enable a judgement to be made on the effect which the plant will have on the community. This includes economic, environmental and other amenity benefits or disadvantages. It would appear from various public enquiries and other studies that the particular social and legislative environment within the UK has enabled risk analysis to contribute to the public debate.

In other countries of Europe different cultural attitudes and types of legislation could well lead to a more absolutist approach by government authorities, requiring strict quantitative criteria to be met.

Where a large number of licensing and zoning decisions have to be made by local authorities, the guidelines for the acceptability of risk need a firm basis which will stand up in political debate and in a court of law.

A study into the methods used for the control of industrial risks in Europe has recently been completed by Pikaar and Seaman for the 'Ministerie van Volkshuisvesting' Ruimtelijke Ordening en Milieubeheer (VROM) in Holland[6]. The survey reviewed practices in France, Germany, Norway, Switzerland, UK and USA and involved contact with over 72 experts from government authorities, industry and research institutes. One interesting conclusion from this study was that:

'The consistency of approach between the major process industry operators is much greater than between legislators; apart from German industries. Major operators performed risk assessment of some kind on selected installations in furtherance of their own safety goals rather as a result of regulatory requirements. There appears to be a broad acceptance by the major process industries that safety decision-making needs to be supported by risk management and that it is not meaningful to do this without a target of some kind.'

5.4   TOLERABILITY AND ACCEPTABILITY OF RISK

It is important to understand that acceptance of an activity should not be based on risk alone. This could lead to automatic acceptance of proposals that just meet the criteria, and rejection of proposals that narrowly fail to meet it. Sound decisions are unlikely to be reached if no consideration is given to the uncertainties in the risk estimates, the cost of reducing the risks, other costs to society of the activity or the benefits to be derived from it. But it can be useful to set design targets and there are levels of risk which society would not wish to see prevail, irrespective of the costs of reduction or the benefits obtained.

The relative significance of quantified risk estimates can be assessed by comparison with other risks that people experience in everyday life. It is recognized that an individual's acceptance of risks is conditioned not just by their chance of occurrence but by many complex factors. However, it is possible to specify, in order of magnitude terms, levels at which an additional risk imposed on an individual must be considered unacceptable, because it would exceed levels of risk associated with activities undertaken voluntarily . At this level the risk may begin to contribute significantly to the average overall individual risk of death in any one year from all causes. It would be difficult to justify the imposition of such a level of risk, whatever the benefit derived.

It is also possible to specify, again in order of magnitude terms, levels of risk to individuals which can be considered to be insignificant as they are very much lower than the average risk of death from accidents and are of similar order of magnitude to events over which the individual has no control whatsoever.

These upper and lower risk boundaries identify the general range of two or three orders of magnitude in which risk should be carefully considered in conjunction with many other features. These may include environmental, employment and commercial considerations. This three-zone approach has been suggested as a decision-making framework in considering the impact of potential accidents on society as a whole. Following the public inquiry into the building of the nuclear power station Sizewell B, the UK HSE issued the report *The Tolerability of Risk from Nuclear Power Stations*[7]. Although the report concentrates on nuclear power it has influenced the development of risk assessment in other fields.

The report defines concepts (see Figure 5.1, page 92) of:
- an 'intolerable' level of risk at or above which immediate action to reduce the risk or terminate the activity is called for, irrepective of cost;
- a 'broadly accepatble' level at or below which further reduction measures are not required;
- a middle region where additional risk reduction measures are necessary until their overall cost becomes grossly disproportionate to the risk reduction produced. This is classed 'as low as reasonably practicable (ALARP)'.

In weighing the costs of extra safety measures, the principle of ALARP applies in such a way that the higher and more unacceptable a risk is, the more proportionately employers are expected to spend to reduce it. At a point just below the limit of tolerability employers are, in fact, expected to spend up to the point where further expenditure would be grossly disproportionate to the risk reduction achieved. As by definition the risk will be substantial, this implies a considerable effort even to achieve a marginal reduction. There may, however, come a point where — with existing technology even a marginal further reduction would be unjustifiably expensive and at that point the obligation to do better is discharged.

Where the risks are less significant, the less proportionately it is worth spending to reduce the risk. At the lower limit where the risks are 'broadly acceptable' the levels of risk are so insignificant that further reduction is not necessary, provided that the risk levels will be attained in practice. The 'broadly acceptable' levels of risk become truly negligible in comparison with other risks to which the individual and society are subjected.

It is important to distinguish between ALARP and ALARA (as low as reasonably achievable). ALARP implies that, in making a judgement, the total cost and inconvenience associated with risk reduction measure may be weighed against the benefits of reduced risk. ALARA implies a stricter test of technical feasibility in determining which risk reduction measures should be adopted.

Further reports have been issued by the HSE, *Risk Criteria for Land Use Planning in the Vicinity of Major Industrial Hazards*[8] and *Major Hazard Aspects of the Transport of Dangerous Substances*[9]. The first report provides guidance on the way in which the HSE assesses applications for developments such as residential housing, or retail outlets, close to major hazard installations. It includes consideration of the treatment of vulnerable sections of the populations such as the young, old and infirm. The report does not directly address the problems associated with the construction of a new major hazard installation. The second report is considered in more detail in Chapter 6.

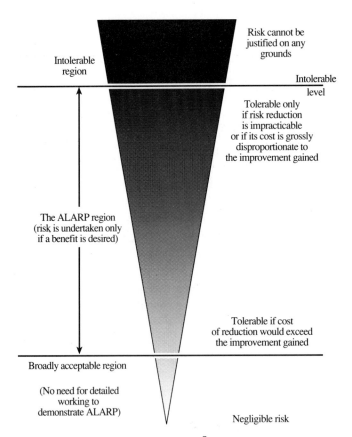

Figure 5.1   Levels of risk and ALARP[7]. (Crown copyright is reproduced with the permission of the Controller of HMSO.)

5.5 COST-BENEFIT ANALYSIS

As noted already, once a risk is found to be in the ALARP region, cost-benefit analysis may be used as an aid to decision making. This may take a number of different forms. In the simplest case, the total cost of the measures necessary to reduce the risk may be set against the achieved risk reduction and decisions made on the best option to adopt.

In other cases, the use of cost-benefit analysis may be extended to the allocation of a notional monetary value to the loss of life (the cost of a life). Techniques of this type have been used for a number of years in the field of road safety in determining which road improvement schemes should be implemented[10]. Despite this there may still be circumstances where the techniques can give rise to fierce debate. In using these techniques, consideration must be given to the differences between the risk to employees and to members of the public. Employees have some element of choice in accepting a risk and can be seen to gain financially through their wages. By contrast members of the public have little choice in accepting a risk and will usually receive no direct financial gain.

Differences in the level of tolerable and acceptable risk for employees and members of the public are now accepted in setting criteria and need to be reflected in cost-benefit analysis.

Care must also be exercised to ensure that every effort is made to identify the most effective way of reducing the risk. Excessive estimates of cost, which then make the risk reduction measures too expensive to adopt, must be avoided. In other cases the use of cost-benefit analysis and the conclusion that the risk reduction measures are too expensive to implement, may reduce or remove the strong pressures on the project team to find simpler, more effective ways of reducing the risk.

Despite these limitations, the use of cost-benefit analysis has increased considerably, especially in the field of transport and offshore safety. In the latter case risks to the public are generally not involved and cost-benefit analysis may be used alongside calculations of the potential loss of life (PLL).

5.6 SOME OTHER GENERAL POINTS

It is inevitable that where a major decision is being made many of those involved, including many of those working in less technical roles for the regulatory authorities, will not have a detailed understanding of risk analysis techniques. This puts a duty on the operators of installations as well as on risk analysts to provide unbiased and understandable information, as responsible members of society.

Experience has shown that full QRA studies can be expensive, although costs are declining and are small in relation to the cost of the installation being considered. The authorities should weigh these costs against the benefits of improved safety and/or better founded decision making when requesting QRA.

QRA studies are best applied where they can really assist with the understanding of a situation which does not have a history long enough to allow justifiable conclusions on the basis of experience. If the studies are too broad and generalized to identify specific problems and the effect of remedial measures, it is improbable that they will lead to effective control measures which could not be achieved by applying established engineering codes of practice, procedures, safety policies and identification techniques.

A QRA study of an installation requires considerable time and effort. This has resulted in the development of short cut correlation methods in order to reduce the work. A major problem of such techniques is that they do not require an understanding of the plant being studied. They assume a constant functional relationship between elements of design and the risk. This is a matter for concern, particularly if the short cut methods involve studies at an early stage in the design, because insufficient information will be available for a full identification of hazards. For an existing plant their use could result in insufficient time being spent on hazard identification. There are many techniques available to perform a rigorous analysis in a cost effective way, and these should be used rather than concentrating on a too limited set of selected parts of an installation, which brings in an element of judgement that has in a number of cases later proved to be unwarranted.

The maximum credible accident approach for assessing the acceptability of an installation has disadvantages. It can lead to far too large an emphasis being placed upon very low probability events, although the possible consequences may be great. It may be thought necessary to consider these from an insurance point of view or when making emergency plans, but in practice these situations are likely to constitute a very low residual risk. It can also be very difficult to obtain agreement between different groups of analysts as to what constitutes a maximum credible event.

When carrying out a QRA it is necessary to verify that the plant or system is or will be constructed, operated and maintained in a competent manner. In addition, the documentation on which the study is based must be correct. It is also necessary to verify that the safety management of the site is up to the industry standard. It is pointless to indulge in the sophistication of a QRA if a plant is badly designed or managed. Such a plant is prone to serious accidents that can be identified and prevented by the use of conventional methods. It is important to remember that risk analysis is only one of the tools in industry's 'safety toolbox'. This aspect of the subject has been covered in the CONCAWE report issued in 1982[11].

5.7   THE WAY FORWARD

The resources for improving safety to meet the economic and social benefits expected by society are limited. Application of QRA techniques can help industry to apply its resources where they can do the most good. However, this objective will not be realized if the techniques are extended beyond the point of credibility of the results to yield unrealistically low probabilities. In 1976 the UK Health and Safety Commission (HSC) Advisory Committee on Major Hazards[12] suggested that the probability of a major undesired event on a plant of once in 10,000 years was on the borderline of acceptability. It is of interest to note that this was the frequency used for deciding upon the height of dykes in the Netherlands. The Norwegian authorities used the a figure of 9.0×10^{-4} per annum in the regulations for assessing the safety and reliability of conceptual designs for offshore construction, but now require companies to develop suitable criteria for their own facilities. In 1986 the Dutch authorities published quantitative risk criteria for both individual and societal risk[13].

A number of public studies have been reviewed by Lans and Bjordal[14]. This analysis supports the view that a process plant which has been designed to correct engineering standards, and which is properly operated and maintained, is unlikely to experience major undesired events more frequently than once in 10,000 years (10^{-4} per year). Nevertheless, some events with a very low probability may potentially have consequences that still warrant their investigation. Pasman[15] has recently reviewed the current understanding of risk perception and of criteria for tolerability. He discusses some of the problems associated with societal risk criteria, particularly when applied to installations with significant value to the community.

Before applying QRA to the more standard type of process plant for the purposes of granting licences or planning permission, the authorities should ensure that the design, construction, operation and maintenance of the plant are up to adequate standards. This responsibility already comes within the sphere of the regulatory bodies in most countries.

Quantification of risk may support safety decisions about plants involving step changes of scale, complexity or technology. The need for this should become apparent when the initial process safety analysis is carried out. The uncertainties involved in the methodology indicate that the contribution of the resultant figures to decision making should be by discussion between knowledgeable members of industry and the regulatory body. These discussions could lead to setting general criteria to be used in routine decision making about licensing and siting. The existence of such criteria should not, however, prevent further discussion especially in borderline cases.

Recent and continuing developments in the process industries, which have their counterparts in other parts of society, are also influencing the development of QRA. Examples are:
- computer technology;
- large, integrated process plant complexes;
- the higher education requirements that are needed for people handling these new systems.

These will have an influence on the development of methodologies for analysing risk situations for people and the environment in which they live. Areas in which change can be expected are:
- a combination of reliability and risk studies with more multi-disciplinary analytical methods, including long-term toxicological effects;
- machine/operator relationships (for example, ergonomics and training);
- better exchange of data via data banks (computer networks);
- better understanding of ways of dealing with uncertainties, where there is lack of knowledge and data (Bayesian estimation, fuzzy set and possibility theory);
- a combination of QRA and cost-benefit analysis as part of the decision process.

The position of QRA in the future will depend not only on how well the analytical methods can give the answers which are needed but also on their cost and comprehension. An important aspect will be their development into a form in which the results can be meaningfully communicated from the analysts to others such as managers, politicians and the public.

Industry and governments have recognized the need to establish more formal procedures for carrying out QRA studies which embody quality principles. The aim is to make QRA studies transparent to their readership and traceable for other analysts trying to verify assumptions or update the study. At the time of writing, a draft international standard developed by the IEC is out for final comment addressing these issues.

Even with these developments, QRA will remain only one part of the total safety package. The primary requirements for safe process plant will always be good engineering, well qualified personnel and good management.

REFERENCES IN CHAPTER 5
1. Health and Safety Executive, 1978, *Canvey: An Investigation* (HMSO).
2. Health and Safety Executive, 1981, *Canvey: A Second Report* (HMSO).

3. COVO Comittee, Rijnmond area, 1981, *Risk Analysis of Six Potentially Hazardous Industrial Objects in the Rijnmond Area: A Pilot Study* (Reidel, Doredrecht).
4. *LPG Integraalstudie*, 1983 (TNO, The Netherlands).
5. *Evaluatie van externe veiligheidsrapporten, number 1992/3* (Ministry of Housing Physical Planning and Environment, dept SVS, The Netherlands).
6. Pikaar, M.J. and Seaman, M.A., 1995, *A Review of Risk Control, publication no. 1995/27 A* (VROM, The Netherlands).
7. Health and Safety Executive, 1992, *The Tolerability of Risk from Nuclear Power Stations* (HSE Books).
8. Health and Safety Executive, 1989, *Risk Criteria for Land Use Planning in the Vicinity of Major Industrial Hazards* (HSE Books).
9. Health and Safety Executive, 1991, *Major Hazard Aspects of the Transport of Dangerous Substances* (HSE Books).
10. Jones-Lee, M. and Loomes, G., 1994, Towards a willingness to pay based value of underground safety, *Journal of Transprt Economics and Policy*, 28, 83–98.
11. *Methodologies for Hazard Analysis and Risk Assessment in the Petroleum Refining and Storage Industry, report no. 10/82*, 1982 (CONCAWE).
12. Health and Safety Commission, 1976, *Advisory Committee on Major Hazards. First Report* (HMSO).
13. *Premises for Risk Management.* Second Chamber of the State General, The Netherlands, Session 1988-89, 21137 nr 1-2.
14. Lans, H.J.D. and Bjordal, E.N., 1983, Application of risk analysis techniques, *4th International Symposium on Loss Prevention and Safety Promotion in the Process Industries. Harrogate 12–16 September*(IChemE), G46–G55.
15. Pasman, H.J., 1993, Risk perception and acceptable risk influencing factors, *2nd World Conference on Safety Science, Budapest, November.*

6. SPECIAL TOPICS IN RISK ASSESSMENT

The preceding chapters have attempted to provide a balanced view of the application of risk assessment in the process industries from the identification of hazards, the calculation of consequences, the quantification of event probabilities and risk, to the use of the results. As can be expected in a field as important as risk assessment, the techniques continue to be developed. This section gives an overview of some of the applications of risk assessment which have emerged recently. They include offshore operation, transport, safety management systems, warehouse storage, effects on the environment and the use of computers in plant operation.

These techniques are in different stages of development. Some, such as the application of QRA offshore, are already well established. Others are still in the exploratory stage. It should be stressed that although these techniques are included here, their application is not recommended in any specific case; rather, this decision must be made by those responsible for the problem. As with other chapters in this book an overview of the application is provided, together with references, which should enable the reader to make an informed decision about whether the technique described is appropriate for the problem being considered.

6.1   OFFSHORE QRA

The use of QRA techniques to assess offshore facilities has grown to represent one of the largest application areas of QRA methodology. These facilities cover a very wide range of designs, including fixed platforms, semi-submersibles, jack-ups, mobile drilling units, and subsea production facilities with moored tankers. Operating phases include exploration drilling, workover, production, simultaneous drilling and production and/or construction and production, and shutdown.

While it might be thought that the QRA of such facilities would be broadly similar to onshore studies, given the similarity of hydrocarbon fluids and process equipment, there are important differences. Important features leading to a different emphasis in the QRA include:
- very small physical area, with a high degree of congestion;
- three-dimensional interrelationships versus predominantly two-dimensional onshore;

- a real possibility of total loss of personnel, production, investment;
- limited range of materials (hydrocarbons, water, hydrogen sulphide, carbon dioxide, etc);
- multiple operational phases over lifetime (drilling, phased production);
- very high pressures and potential flowrates;
- massive inventories (well blowout or potentially over 100 km of pipeline fluids);
- only employees and contractors at risk, no general public exposure;
- close proximity of wellheads, production, cranes, helicopters, accommodation;
- heavy capital investment with significant reliability and availability concerns.

These features lead to studies with the following characteristics when compared with onshore QRA studies:
- there is wider selection of events under consideration (for example, blowouts, structural failures, helicopter incidents);
- there is greater focus on escalation events;
- there is greater focus on protective systems;
- there is greater focus on emergency evacuation of personnel.

6.1.1 HISTORY

The two most important events in European offshore history have been the capsize and sinking of the Alexander L Kielland (1980) and the fire and explosion aboard the Piper Alpha platform (1988), in the Norwegian and UK sectors of North Sea, respectively. Both of these total loss events led directly to regulations requiring quantitative evaluations of safety characteristics, initially in the Norwegian sector and subsequently in the UK sector. There have been several other total loss events around the world.

The Kielland disaster on 27 March 1980 involved the catastrophic collapse of one of the five legs of a floating accommodation platform. The design did not consider such an event adequately and the actual capsize time of 20 minutes was much shorter than had been anticipated. Many of the escape and evacuation systems did not work properly. A total of 123 fatalities occurred, with only 89 people rescued.

The first formal introduction of QRA techniques in the offshore industry was instituted by the Norwegian Petroleum Directorate with its *Guidelines for Concept Safety Evaluation (CSE) Studies* in 1981. Elf's Heimdal field was the first to comply in 1982. The CSE regulations required nine separate types of incident to be evaluated with respect to three safety functions for risk and to demonstrate that each had a failure likelihood of 1×10^{-4} per annum or less. The safety functions related to temporary refuge, escapeways and means of evacuation (for example, lifeboats). The quantitative criteria were treated flexibly, with many operators setting a total criterion of about 1×10^{-3} per annum. The intent of the Norwegian regulations might be interpreted as being directed as much to maintaining the integrity of the installation as to protecting life, but this would be a simplification. Lives were protected by protecting the integrity of escape routes, safe havens and the structure of the installation itself. More recently, the Norwegian regulations have been updated to account for experience, and in several respects they have converged with the UK regulations; both now require risk analysis and formal safety management systems.

The Piper Alpha platform suffered a blast and pool fire event on 6 July 1988, which in principle should have been survivable, but which led to the rupture of a main riser pipeline, and when that failed there was a massive escalation which led ultimately to 167 fatalities and the total loss of the platform.

A thorough and far-reaching Public Inquiry was chaired by Lord Cullen. He recommended that the operator of an offshore installation should develop a safety case and, within this, demonstrate using QRA techniques that acceptance standards were met with respect to personnel protection facilities. Control was transferred to the HSE, which was already responsible for most areas of safety, and it implemented the Cullen recommendations in the subsequent Offshore Installations (Safety Case) Regulations of 1992.

These regulations are a good reflection of current thinking on the contents of offshore safety cases. They require that all potential major accident hazards be identified, the risk of these hazards evaluated, and that suitable measures be implemented to reduce risk to people to a level regarded as ALARP.

The HSE has identified what it sees as the primary objectives of QRA:
- to provide insight into possible incidents, their consequences and frequencies;
- to establish the possible escalations from initiating incidents and the factors contributing to these;
- to determine the risk ranking of the major contributing events;
- to demonstrate the degree of risk and whether this meets corporate or regulatory targets for ALARP;
- to assess the risk reduction potential of a suitable range of mitigation measures.

The safety case regulations provide a list of nine specific hazards to consider:

(1) Hydrocarbon releases leading to fire, explosion or toxic effect arising from any part of the drilling, processing, import or export facilities.
(2) Vessel impact including: support vessels, crane barges, fishing boats, bulk oil transports and loose single-buoy mooring vessels.
(3) Structural failure or foundation subsidence, extreme wind or wave, seismic event or ice loading.
(4) Helicopter or fixed-wing aircraft impact.

SPECIAL TOPICS IN RISK ASSESSMENT

(5) Dropped objects or loads.

(6) Non-process fires in accommodation, electrical equipment, and fuel storage.

(7) For mobile or floating units: loss of station keeping, loss of stability or loss of buoyancy.

(8) Effects from nearby installations (for example, connected by pipelines, oil slicks).

(9) Hazardous activities (for example, diving, crane operations, construction accidents).

6.1.2   DETAILED ANALYSIS

Most of the techniques described in this section have been developed for the UK and Norwegian sectors of the North Sea, and are starting to find acceptance worldwide (for example, in Australia, Canada, Malaysia and Brunei).

Blowout events

Well blowout events are usually treated separately from process events as these are fed directly from the reservoir. Several different types of well phases can be identified, each with different hazards. These include exploration (including appraisal) wells which have a drilling phase and development wells which have drilling, completion, production and workover phases. The hazard of drilling into shallow gas pockets is normally modelled separately, as key features assisting in regaining control (the first casing and the blowout preventer (BOP)*) may not be installed.

Most often historical data is used to analyse such events. Good data is available for the North Sea and the Gulf of Mexico and this is available in databases such as WOAD (see page 72).

Riser and pipeline accidents

Riser and pipeline accidents occur in sea bottom pipelines and vertical risers which carry product from other platforms or export products from that platform. They often contain large inventories at high pressure and several large accidents (for example, Piper Alpha) have been associated with their failure. While the need for emergency isolation of these lines is clear, the location of such isolation is more problematic. It might be on the platform, in the splash zone (that is,

---

*   The first casing is the topmost containment sleeve of the well, and the BOP is the valve arrangement at the well exit used for shut in purposes.

101

below deck above the waterline) or on the bottom (the so-called SSIV — sub sea isolation valve). The cost and reliability implications of where to locate an SSIV were serious and this type of analysis formed an early focus of several QRA studies. The findings of these studies contradicted initial views that subsea isolation was always desirable. In fact it was found that the problem was platform specific; there was a tradeoff in diver maintenance risks, vulnerability, cost and reliability issues.

Safety studies should ideally address many factors of importance to the riser/pipeline systems, including location, material of construction, physical protection, etc. Data have been produced recently by Advanced Mechanics and Engineering[1] on behalf of UK Offshore Operators Association (UKOOA).

Process releases

QRA of offshore process releases differs from onshore risk assessment in both escalation modelling and outcomes. Causes of process releases include impacts (dropped or sliding loads), corrosion (internal or external), erosion, mechanical defects, operational errors and natural hazards (mainly wind). Several approaches are used for identifying releases. The most rigorous selects three to five leak sizes for every isolatable section. Leak sizes may be defined in terms of hole size or release rate. An approach based on hole sizes, as is usually adopted for onshore studies, might be regarded as more intuitive. Inside a confined module, however, most releases of a given rate result in similar outcomes regardless of their specific location within the module, allowing all similar releases to be clustered for analysis purposes. Thus an approach based on release rates reduces the number of computations required.

Once a release case is established, a full event tree should be determined by the analyst to establish what possible outcomes may result. This is the detailed escalation analysis that typically characterizes an offshore QRA. Much of the analysis effort is devoted to this area.

Factors of importance in the escalation analysis include:
- release orientation;
- relationship to other equipment items, firewalls, escapeways;
- human response;
- leak detection;
- valve isolation;
- actuation and blowdown;
- deluge or other active fire protection systems;
- passive fire protection systems;
- wind direction and effect on module ventilation;
- ignition and ignition delay.

Offshore QRA studies tend to model the effects of safety features carefully, as a key use of such studies is to establish the costs and benefits of a range of such measures.

The event tree analysis is an area of major divergence from onshore QRAs. It is common for onshore QRAs to employ relatively standardized event trees (for example, SAFETI, RISKAT) with little analyst effort required, whereas in offshore QRA these are platform specific and typically become the largest single QRA task. The more detailed analysis also has implications for the failure frequencies ascribed to items. Onshore generic failure frequencies tend to be escalated data — that is, if a 100 mm leak started as a 10 mm leak but escalated to 100 mm it will tend to be recorded as a 100 mm event. Offshore, that increase would usually be considered in the escalation analysis. (The reason for this difference is that there have been no onshore data collection exercises as detailed as those carried out in the North Sea by the HSE and industry in line with the recommendations of the Cullen inquiry. Thus data for the same equipment *may be different and should not be used interchangeably*, unless the underlying nature of the data and its intended application are understood clearly and the analysis modified accordingly.

Non-process fires
Consideration of accommodation, electrical, and non-process flammable materials (eg. aviation fuel, diesel, methanol, etc) is necessary, but these will often be of minor importance.

Ship collision
Ship collision is a unique danger for offshore platforms. The most serious result can be leg failure leading to catastrophic collapse, but a range of other serious events can occur, such as riser failures. Collisions are typically divided into those associated with errant powered vessels (for example, those bound for the installation or those navigating nearby) and drifting collisions (for example, drifting vessels or anchor chain ruptures). Data are available from historic accidents to estimate likely collision energies and effects for a range of vessel sizes. Commercial and fishing vessel movements can be combined with installation-specific traffic to establish a likely collision frequency. Well-known codes for this type of analysis include CRASH and COLLIDE.

Helicopter crashes
Helicopter crashes can often be characterized in terms of accidents in-flight and those during take-off or landing. In-flight accidents affect only the passengers

and flight crew, while the latter also affects platform personnel. Mostly, analysis of these events is by reference to generic historical data. This data are made platform-specific by reference to the distance flown, helicopter types, passenger loads and flight frequencies.

An interesting factor in retrofitting safety features is the temporary increase in numbers of contractors necessary to implement the measure. This can lead to an increase in helicopter-related risk greater than the risk reduction provided by the measure itself, thereby negating the benefits predicted.

Crew boat accidents
QRA for crew boat accidents is analogous to helicopter accidents, but applies to installations served by crew boats, supply vessels and standby vessels. The analysis would address sinking or collision leading to passenger fatalities and collisions leading to installation effects.

Workplace accidents
Other events, often termed workplace accidents, cover a range of non-process situations such as falls, loads dropped onto people, asphyxiation, bridge collapses and diving accidents. Such events are not usually modelled; instead historical data are applied, often with factors to allow for features on the installation or for different worker groups.

Structural failures
Structural failures can be associated with several cause mechanisms — for example, extreme weather (wind or wave), earthquake, loss of buoyancy, excess weight, design fault or fatigue. The offshore accident database WOAD shows eight total losses worldwide of fixed offshore structures in an 18 year period due to weather-related events. The failure rate of structures designed after 1970 is much less than earlier structures due to a change in design standard; they are now designed to withstand a 100 year wave rather than a 25 year wave. Very detailed probabilistic structural models are available for comprehensive modelling (for example, SESAM). Such modelling is usually conducted outside the scope of the QRA, but is an integral part of the overall safety review.

Dropped objects
Crane and winch lifting activities are common offshore. Dropped object events are a significant risk, due to the cramped quarters and poor weather conditions. Good historical records have been kept and form the basis of risk predictions. The level of analysis can be matched to the degree of risk expected. Care is

necessary to avoid double counting of process pipe leaks which are similarly based on historical data, and which already include this mechanism.

6.1.3 CALCULATION METHODS

There are many calculation approaches adopted for offshore QRA. With the exception of the structural failure analysis, which typically employs probabilistic methods, most QRA employs deterministic calculations. In some studies, however, the number of discrete event tree outcomes which are simulated can become so large that they approximate probabilistic analyses.

A range of calculation tools have been developed for offshore QRA. They are listed, roughly in order of development, here:

Simple spreadsheets

Simple spreadsheets tend to employ simplified event tree analysis and usually import consequence results from external calculations. The range of analysis undertaken does not always have the discrimination to demonstrate the value of important safety features. Spreadsheets tend to be used for regulatory concept safety evaluation (CSE) analysis as a pass/fail analysis. Commercial spreadsheet packages are usually employed as these are very flexible and easy to change to account for platform-specific features of the analysis. Compiled computer programs tend to be much less flexible in addressing these necessary differences.

Detailed templates

The detailed template approach recognizes that there are enough common features between platforms to allow the use of a more standardized approach, and adequate accuracy can be achieved at lower cost. This standardized approach is of more value in a pass/fail mode than in a platform-specific risk reduction, and safeguards cost-benefit analysis. Again, commercial spreadsheet packages are employed.

Complex spreadsheets

Complex spreadsheets started to be developed in the late 1980s, reflecting the increased power of modern spreadsheet tools (for example, spreadsheet linking, graphics and macro capabilities). Ramsay *et al*[2] provide a useful review of the state of the art in this area. These models tend to carry out most parts of the analysis inside the spreadsheet with release rate, dispersion and fire and explosion calculations embedded rather than imported. Extensive escalation analysis and semi-automatic logic are employed. These models were able to meet the objectives of the HSE for QRA, in terms of understanding the accident mechanisms and escalation, the assessment of mitigation alternatives and demonstration of ALARP.

These models address in great detail the benefits of safeguards, and allow risk mitigation to be the primary focus of the analysis rather than pass/fail analysis. These spreadsheets invariably become very large and complex and consequently increasingly greater efforts are necessary to verify and audit the tool.

Integrated systems

By 1990, the offshore industry collectively realized that more formal calculation methods were necessary. The problems inherent in spreadsheets (ease of intended and unintended change, difficulty in verifying formulae, lack of structure, etc) meant that important parts of the safety design were being carried out by techniques that did not lend themselves well to quality verification.

Consequently 14 North Sea companies, consultants and governments joined together and sponsored the development of OHRAT — the offshore hazard and risk analysis toolkit[2]. This is best described as a risk modelling environment, rather than a risk calculation tool (like SAFETI or RISKAT for onshore QRA). OHRAT allows the analyst to combine data, models, results and all necessary interlinkages in the form of a windows-based icon map. While a basic set of consequence models and frequency data are included, the design of the environment allows easy importing of any suitable model or data. Audit and verification of the software, and documentation of logic are all much superior to spreadsheet tools.

An alternative QRA approach, also offering the benefits of compiled computer code and auditability, is the PLATO model[3]. PLATO takes a detailed description of the platform (process equipment, piping, firewalls, deluge, etc) and generates an automated event tree analysis. As previously noted, event tree analysis is one of the major activities in offshore QRA, and PLATO uses an initially defined database of escalation assumptions to model all phases of an escalation sequentially (for example, jet leak, ignited, impinges on firewall, firewall fails in 10 minutes, jet now impinges on riser, riser fails).

The PLATO approach can automatically generate thousands of event trees, compared to the dozens or hundreds typical with more manual approaches (spreadsheets or OHRAT). Due to the large number of calculations generated, consequence modelling is simple. There is a trade-off, however, between the benefits of extended automatic event tree analysis on the one hand and the improved appreciation of the specific escalation pathways obtained from manual event analysis on the other.

SPECIAL TOPICS IN RISK ASSESSMENT

6.1.4 RESULTS PRESENTATION

Results for offshore QRA are now most commonly used for risk reduction cost-benefit analyses as part of the risk ALARP demonstration. The risk measures employed tend to be single number measures, making cost-benefit analysis easier. Common measures are the potential loss of life (PLL — expressed as fatalities per year), the FAR (Fatal Accident Rate fatalities per 10^8 exposed hours), and the average individual risk (whole platform or for individual jobs — driller, operator, cook). It is unusual to generate F–N curves and risk contours (see Chapter 4) for offshore QRA, although there may be occasions when these could be useful.

An important factor offshore when assessing the costs and benefits of retrofitting mitigation measures is the construction activity. As production often continues during the installation, the number of people on board increases, and the construction activity itself generates risk. More sophisticated risk presentations include the increase in risk associated with these activities. It is not uncommon to see negative mitigation values when these factors are considered.

An example risk presentation showing the effects of risk reduction measures with and without construction activity is shown in Figure 6.1.

6.1.5 CONCLUSIONS

Offshore QRA differs markedly from onshore QRA, both in the scope of the study and in its detailed application. It tends to be used more in the context of

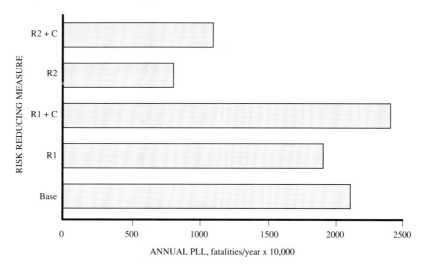

Figure 6.1 Typical results presentation comparing risk reduction measures. Base: base case; R1, R2: remedial measures; C: including construction.

107

flexible ALARP risk reduction alternatives, rather than for more formal onshore land use planning criteria. The focus offshore is thus usually a comparison of risk reducing measures, often with quite detailed event tree modelling to demonstrate the benefits.

A major trend in offshore QRA is the increasing emphasis on quality and auditability of the whole QRA procedure, with a requirement of traceability throughout the whole analysis from the identification stage right through to mitigation analysis. This is likely to provoke a reduction in use of customized spreadsheet tools and greater application of compiled computer codes.

Other trends are an increasing use of risk technology for economic assessments (such as the reliability and redundancy of key systems) and the use of risk escalation analysis to verify the operational response of emergency systems designed to code or other assumptions. Current development projects cover more focused offshore leak data collection, smoke modelling using sophisticated computational fluid dynamics techniques, frequency modelling (for example, making leaks less generic and enhancing ignition modelling) and better communication and analysis of results.

6.2   TRANSPORT RISKS

Hazardous materials are transported in significant quantities by road, rail, barge, vessel and pipeline. There have been a number of transport accidents in Europe and elsewhere (for example, San Carlos tank truck BLEVE, the Bantry Bay oil tanker explosion and the Russian LPG pipeline disaster) which have highlighted the hazards resulting from the transport of dangerous goods. Furthermore, transport systems impose risk implications which are significantly different from stationary facilities.

For example, when comparing road and rail transport with stationary facilities there are both risk-lessening and risk-increasing factors. Risk-lessening factors include:
- transport containers tend to involve relatively small inventories of material;
- cargo temperatures are usually near to ambient;
- pressures are moderate (vapour pressure or padding gas pressure only);
- no chemical reactions or processing is occurring.

Risk increasing factors relative to stationary sources include:
- the dangerous material is brought into close proximity with the general public;
- safety systems carried onboard tend to be fewer and less robust than fixed systems;
- there is no secondary containment, if the primary containment is lost;
- the driver (and mate if provided) are the only immediately knowledgeable people;

- site-specific emergency response plans cannot easily be developed or communicated;
- the transport container is subject to many external threats;
- transfer operations mean that connections are frequently made and broken.

Some problems faced when setting criteria for transport risks include the fact that many people are exposed to the risk along the whole route, but their individual exposure is of short duration. Individual risk (which is easier to understand) may therefore be poor indicator compared with societal risk. Societal risk in principle is a superior means to assess transport risks, but it poses certain unsolved practical difficulties. The most difficult is that societal risk increases roughly linearly with route distance, thus setting a readily understandable transport criterion that has not been solved. Finally, a large process facility brings local benefits (taxes, jobs, services), whereas transport tends to bring few benefits and many perceived problems to those locally affected. All these factors make for difficulties in generating well-founded transport risk criteria, and greater problems are being encountered in defining these than in the actual calculation of the basic risk result.

Pipeline transport, which might be considered to involve a stationary system, introduces special hazards of its own. These include large inventories and flowrates with less isolation opportunities, greater exposure to third party digging activities, difficult inspection and close approach to populated areas.

In general, the transportation of dangerous substances is subject to national and international regulations. Some companies and countries concerned at the risk have carried out QRAs. These studies range from the early US Coast Guard Vulnerability Model (1970s), through the Netherlands Integrated Study on fuel risk alternatives (1980s), to the UK Transport Risk Study (1990s). The UK Transport Risk Study is considered a good example to demonstrate the current approach to transport risks. Risk techniques are also being more widely applied to transport issues not involving dangerous goods (railway safety, tunnel safety, airport runway positioning and air traffic route separation, etc), but discussion of these is beyond the scope of this book.

6.2.1 UK TRANSPORT RISK STUDY

The UK HSC Advisory Committee on Major Hazards[4] identified a prima-facie case for further investigation of the risks associated with the transport of hazardous substances. In response, HSC set up a subcommittee of the Advisory Committee on Dangerous Substances (ACDS).

Initially, the ACDS subcommittee attempted to approach their task using current legislative standards and their engineering underpinning as bases for their investigations. It soon became clear, however, that the historical record was quite insufficient to provide a robust basis for prediction of major events of low probability, and the use of wider data sources (for example, international statistics) was fraught with difficulties of relevance and extrapolation. International transport controls are well-known and highly developed — for example, rail (RID), road (ADR) and marine (IMDG), at the core of which lie the United Nations unifying concepts. But even with strict enforcement these leave a residual risk, which may not adequately address very serious, rare events. These and other problems led to a decision to proceed on the basis of a systematic QRA. This decision was tempered, however, by a recognition of the inherent uncertainties in current QRA.

The ACDS[5] study in 1991 looked at both marine shipping risks and road and rail surface transport; but in order to limit the scope, only the most dangerous goods were addressed. For road and rail these included liquefied, toxic and flammable gases (with LPG, chlorine and ammonia as examples), large volume dangerous goods (motorspirit) and explosives.

The QRA procedures used followed a classical risk assessment procedure, essentially based on the 'Canvey' and 'Rijnmond' formats, involving :
- review of movements;
- review of historical accident record;
- identification of main or potential causes of classes of incidents;
- estimation of event frequencies (for 'specimen' substances);
- estimation of appropriate 'source term(s)';
- quantification of consequences;
- identification of individual risk levels and societal risk levels;
- identification of remedial and mitigating measures;
- evaluation of remedial and mitigating measures.

The marine assessment differed somewhat from the road and rail study, reflecting the nature of problem being addressed and the special characteristics of each. The marine risk approach involved a two-step procedure for reasons of efficiency. This was due to the large number of ports (42) in Great Britain handling significant quantities of dangerous substances. Three representative ports were chosen for detailed risk analysis using SAFETI, calculating individual and societal risks for port workers, ship crews and members of the public. Details of traffic and other characteristics of the remaining 39 ports were then combined with the outputs of the detailed assessments of the three ports and scaled to derive a national risk figure.

The road and rail study used the HSE RISKAT package for risk calculations. Puncture frequencies were derived using historical evidence where available, and consensus engineering judgment where such history was not available or felt to be misleading. Frequencies of equipment failure and small leaks were similarly derived. The estimated spill probabilities were combined with typical population densities, weather data and vulnerability/effect models to estimate risk to those living adjacent to the road and rail routes, rail passengers *en route*, and road users. Representative routes were assessed for each trade and transport mode. This approach permitted the national road or rail trade in that substance to be calculated, by scaling to total national trade and to average route length.

The risk levels predicted were generally found to be low for the transport alternatives considered. The most prominent contributors (within a very small total) were:
- in ports, cold rupture and ignition of consequential spill;
- in ports, unloading point releases;
- for roads, motor spirit (for low consequence results);
- for roads, LPG (for high consequence results);
- for rail, aspects of ammonia transport;
- for explosives, road and rail, initiation by fire.

The assessments did not produce any risk results which, when crudely converted to annual expectation values, predicted more than two deaths per year; often in specific situations the prediction was much lower. Such figures need to be considered alongside the direct transport risks associated with straightforward vehicle accidents. Following a suggestion by the adhoc criteria group to put the value of a life at £2m, the scope for improvements was seen to be very limited. It was in the context of these constraints that specific measures for risk reduction were assessed.

6.2.2   PRINCIPLES OF RISK CONTROL AND RISK CRITERIA
The general principles associated with risk control in the ACDS study used a 'three-zone' approach as described in Chapter 5.

The ACDS[5] report discusses at length the complexities in deriving tolerability criteria for aspects of societal risk, and in the light of all the problems involved suggests criteria for a three-zone approach for ports and for road and rail risks. They apply to one locality or to one route or to the transportation of one substance in the context of a major national transportation activity. They may not be appropriate criteria for — for example — different types or sources of risk, or for judgments about single installations, or for control of land use and development. Higher criteria may apply to the total national port risk(s). A fuller description of this national scrutiny method is provided in the ACDS report[5]. Such essentially localized criteria do not present the full picture for road and rail risks because of the *en route* dimension. For this reason the overall road and rail risks were set against the national port scrutiny level and against local tolerability levels (a more certain test, since if this criterion is not reached the risks must be in the tolerable region).

6.2.3 OTHER TRANSPORT ACTIVITIES

In 1992 the First International Consensus Conference on the Risks of Transporting Dangerous Goods was held in Toronto to provide an open forum for the comparison of risk assessment models and to promote discussion of the dangerous goods transport issues upon which many important public policies are based. The event was structured around a 'corridor exercise' which involved a hypothetical dangerous goods route which the participating organizations were invited to assess prior to the conference, at which the results would be presented[6]. While many details of the corridor exercise generated debate, consensus on sources of variability in the risk result could not be achieved. This led to a call for a code of practice in the area of dangerous goods transport risk analysis which would recommend:

- definition of all assumptions made in the analysis;
- use of standard reporting formats and measures of risk;
- the provision of full documentation;
- internal consistency (intermediate results should be verifiable);
- clear statement of the uncertainties;
- sensitivity analysis;
- advice on mitigation;
- provision of a non-technical communication of risks.

6.2.4 CONCLUSIONS

New and developing technologies require difficult judgements to be made, at various levels and in various areas of society. This requires a transparent decision and risk communication process. QRA provides elements of such a process, providing a procedure, often rigorous, which may involve

- quantification of likely risk with an understanding of the inherent uncertainties;
- reference to the benefits generated by the process, and the political and economic considerations associated with it;
- judgements about 'tolerability' or 'acceptability' for groups directly or indirectly affected;
- decisions as to further reductions in risk, taking cost (including effort) into account.

Greater use of risk-based assessments is expected in the transport area, but problems relating to consistency and suitable criteria remain to be solved.

6.3 SAFETY MANAGEMENT SYSTEMS

During the 1980s there were a number of disasters, some within the process industries, which increased the emphasis on the development and maintenance of effective safety management systems. A number of guides have been developed based on engineering judgement to define necessary elements of a safety management system. Examples include *API RP750 : Process Safety Management*[7], the *CCPS Guidelines on Technical Management of Process Safety*[8], and the *DNV International Safety Rating System*[9]. Recently more fundamental research has been carried out[10].

Formal enquiries into major accidents have pointed to a wide range of factors which may have contributed to, or caused, the accident. Such causes are often described as either immediate or underlying. This makes the distinction between such errors as the opening of an incorrect valve (an immediate cause of a loss of containment incident) and the underlying cause of the incident such as the poor design of the plant. The underlying cause makes the immediate cause highly probable. As already discussed in Chapter 4, human error plays a major part in accident causation.

Here the distinction between human error, either as an immediate cause of the failure or as an underlying cause, is important to appreciate. Generally, the attribution of human error by an individual as the cause of an accident is no longer accepted as giving a full and adequate account of the whole range of causes, nor is 'human error' considered to be unavoidable. Rather than being the main instigators of an accident, operators are often the inheritors of a situation created by poor design, incorrect installation, inadequate procedures or poor management decisions.

These developments have led to is a widespread acceptance of the concept that the way safety is managed on a site directly affects the safety performance of that site (a point already stressed in this book). The acceptance of the importance of a good safety management system (SMS) has led in turn to the development of audit methods which can be used to assess the quality of the management of safety. Safety management audits are not to be confused with the normal monitoring which a site management uses daily to control operations on the site. Nor should they be confused with technical audits which aim mainly to address 'hardware' issues. An audit of the SMS on a plant is normally carried out externally to the management systems, either by staff from within the company who have this independent function or by external consultants or regulators.

The HSE publication, *Successful Health and Safety Management*[11] describes the process of management control and distinguishes between the roles of monitoring and auditing (Figure 6.2 page 114). There are a number of safety management system audits in use in the process industries and common features

113

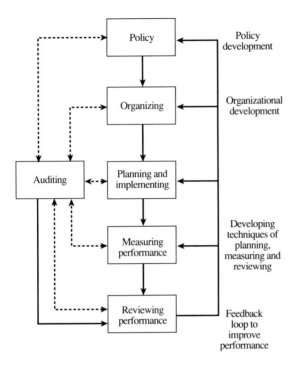

Figure 6.2 Key elements of successful health and safety management[11]. (Crown copyright is reproduced with the permission of the Controller of HMSO.)

of these audits may be identified. There is a wide measure of support for the use of these audit systems as a way of ensuring that programmes of improvement are set in train against a measured base line of performance. DNV's International Safety Rating System (ISRS) has been widely used in the process industries for this purpose.

6.3.1 SAFETY MANAGEMENT SYSTEMS AND QRA

It is against this background that work has been undertaken in recent years to try and improve understanding of the relationship between QRA and SMS. Since QRA provides an assessment of plant safety, and safety in turn is highly dependent on safety management, the link between QRA and SMS is important. It is not uncommon for a consideration of the QRA of a plant and the performance of its SMS to be kept quite separate because. while the interrelationship between them is well appreciated, it is not easy to quantify.

QRA as applied in the process industries often makes use of 'generic' failure rates for loss of containment from items of plant such as pipework and vessels. In this context failure rate data is derived from both historical and theoretical data and may contain elements of engineering judgement and caution. A single 'generic' value is taken to be representative of this data. There is a growing awareness that this generic data may not be appropriate to use in all circumstances, unless the role of management and the organizational factors relevant to the data is identified and made explicit.

6.3.2 QUANTIFYING THE LINKS BETWEEN SMS AND QRA

Thus the past ten years or so have seen the development of methods to describe and quantify the links between SMS and QRA. Three methods have been developed in the UK and The Netherlands. Although they have different origins, and experience of their use is very different, they are essentially similar in their overall approach.

The methods are:
- The instantaneous fractional annual Loss (IFAL) method and the management factor technique.
- Management assessment guidelines in the evaluation of Risk (MANAGER)[12,13].
- The sociotechnical audit method[14,15].

Of these, only the second and third have developed to the point where they can be applied.

These methods involve the calculation of risk using a 'hardware only' QRA, followed by a modification of the risk calculation based on a site-specific audit. That is to say, QRA is used to calculate the risk using details of chemical inventories and process details together with generic failure rate data in the way described in Chapter 4. Because generic failure rates are used which assume a 'baseline' level of safety management, it may generally be assumed that a QRA using generic failure rates is representative of at least an 'acceptable' level of safety management. The risk figure is then modified on the basis of the assessed standard of management on the site. This produces a QRA which is site-specific and incorporates the perceived standards of the SMS at the site.

The MANAGER system was early pioneering work and has been applied widely and experience with the system described[13]. It gives equal weighting to each question in the question-set, and uses a quantification process which is largely based on expert judgement. It provides a method which can, with reasonable time and resources, give a snapshot of the performance of the SMS and a set of recommendations for improvement. Priorities can be put against these recommendations based on the qualitative evaluation of their significance.

Application of MANAGER on over 50 sites has produced findings indicating that the quality of the SMS may reduce risk estimates based on generic failure rate data by about half an order of magnitude, or increase them by about one order of magnitude. These quantitative results have been confirmed on one paired comparison of two similar facilities operated using different SMSs, but this is not formal validation. The technique offers an alternative to other methods, such as the international safety rating system (ISRS), which do not allow the auditors' assessment to be used in any risk assessment carried out at the plant.

The sociotechnical audit method has involved extensive research and in-depth analysis of pipe work and vessel failure data. This provides both a theoretical model to structure the audit question-set, and a statistical basis to weight the audit areas and quantify the range of plant failure rate data[14]. It indicates a range of one order of magnitude for plant failure rate data.

The audit question-set can also be used as a stand-alone audit with a sound theoretical and statistical basis to prioritize recommendations for further improvements to the SMS at sites. It is in this context that Ratcliffe has described the STATAS audit system which evolved from this research work[17,18]. There is now a body of experience of use of this methodology, although further development is still required.

Computer systems linking the other elements of risk assessment described in this book with some of the elements of SMSs and audits are under development, although are still at the trial stage[16].

It is clear that there are important links to be made between the quality of the SMS at a plant and any risk assessment carried out at the same site. The work referenced here, and the systems developed, have gone a long way towards establishing these links quantitatively.

One difficulty relates to the durability of the management factor assessment and the speed with which the 'quality' of the SMS may change over time. It is sometimes argued that because of this, an assessment which includes such considerations is liable to random change at short notice which would invalidate the risk assessment and any decisions based on it. But this uncertainty needs to be placed within the context of the other, very considerable, uncertainties to which the QRA process is subject, and the considerable role which engineering judgement plays in the process.

6.3.3 CONCLUSIONS

Standards of safety management remain an important factor in determining risk at a site. Any assessment which fails to examine the effects of such variability may in itself be intrinsically flawed. A close examination of a risk assessment will clearly show these 'fault lines', which may undermine the validity of the decision-making process. Of course, certain minimum standards must be maintained — and be seen to be maintained — if any decision making based on risk assessment is to have lasting credibility. The approaches outlined here allow the assessor to ensure that — as one pivotal element of the assessment — key SMSs are at least of a minimum acceptable standard. They may also allow identification of areas for improvement. In this way the robustness of the QRA-based decision may be ensured.

6.4 CHEMICAL WAREHOUSE STORAGE

Incidents in a number of countries have shown that a major uncontrolled fire in a chemical warehouse can result in significant offsite effects. The products of combustion and smoke may affect a large section of the local population and contaminated fire water may lead to river pollution or to the contamination of soil and ground water. Following the warehouse fire at Sandoz[19] the Seveso directive was amended to include certain chemical warehouses. In addition CEFIC has issued a *Guide to Safe Warehousing to the European Chemical Industry*[20].

The methods available to assess the threat posed to man and the environment from a chemical warehouse fire are still in the early stages of development. The EU has recognized this and programmes to improve the methods available have been included in the Science and Technology for Environmental Protection (STEP) programme. A description of current EU major hazard and industrial fire research is given by Cole and Wicks[21].

At the time of writing a logical qualitative assessment of warehouse fires coupled with semi-quantitative estimates of the consequences seem likely to produce a more useful assessment than one based on full quantification. Such a qualitative/semi-quantitative approach provides an understanding of where the major risks lie and how they may be eliminated or controlled. Elements of the approach include:
- material hazard assessment;
- development of credible fire scenarios;
- assessment of fire prevention/protection measures;
- consequence assessment;
- elimination or mitigation of consequences;
- quantification of risk.

6.4.1 MATERIAL HAZARD ASSESSMENT

A material hazard assessment must take account of the properties of the materials being stored, the method of packaging and the mixture of materials which may be stored. In the design of a new warehouse such an assessment may be included in the early hazard studies.

Information required includes not only the flammability of the materials but the heat sensitivity, reactivity and the toxicology of both the material being stored and the combustion products[22], the total quantity of material and the effectiveness of the fire-fighting measures. Rapid-acting systems, such as automatic sprinklers, can reduce significantly the quantity of water needed to contain a fire.

In order to gauge the potential effect of a chemical on the environment, information may be obtained from the *Liste Wassergefährdener Stoffe*[23] prepared by the Verband der Chemistre Industrie. In addition, reference should be made to the limits imposed by the EU and national governments, and to the guide produced by the Confederation of Fire Protection Associations of Europe[22] and the *Handbook of Environmental Data on Organic Compounds*[24].

6.4.2 CREDIBLE FIRE SCENARIO

Although research work is in hand to develop models of fire spread, well established empirical approaches may be used to provide a realistic estimate of the rate and extent of fire spread. These methods take account of the development stage of the fire through full development to decay. Estimates can be made of the likelihood of structural failure of the warehouse. Burning rates for a number of warehouse fires are given in Theobald[25].

6.4.3 EFFECTIVENESS OF FIRE PREVENTION/PROTECTION MEASURES

Work over many years has provided a good understanding of the effectiveness of both active fire protection measures (sprinklers etc) and passive protection measures (fire cladding of structural elements, office walls etc). This information, primarily developed for non-chemical fires, can be adapted to chemical fires[26,27]. Particular care is needed when sprinkler systems are installed where flammable liquids are stored.

6.4.4 CONSEQUENCE ASSESSMENT

Fire invariably presents a risk to life. The safety of building occupants may generally be assured by designing the warehouse to legislative requirements which specify alarm facilities, emergency escape routes and evacuation procedures.

Only recently has attention been given to the potential effect of the products of combustion. Few methods are available at the time of writing although considerable work is in hand to improve the modelling of smoke dispersion. Some warehouse analysiss have used conventional Gaussian buoyant gas dispersion models. Although the potential risks must be taken seriously, nearly all fires produce toxic combustion products and ill effects are, in general, limited to those very close to the fire. Recent work has studied the effect of fire on complex chemicals[28].

In addition to the effect of smoke, consideration must be given to the fate of contaminated fire water. Many containers will be damaged in a fire and fire-fighting techniques may use considerable quantities of water. It is therefore likely that an aftermath of such fires will be large volumes of fire water contaminated with a wide range of chemicals.

6.4.5 ELIMINATION OR MITIGATION OF CONSEQUENCES

Whatever the system used, provisions for the containment of contaminated fire water is likely to be needed, depending upon the environmental consequences of spillage. Definitions of various classes of environmental incident are being developed by the UK Government's Department of the Environment amongst others.

Quantitative guidelines on water retention capacity have been developed in Germany[29]. These take account of:

- the material being stored;
- fire detection and alarm provisions;
- fixed fire-fighting equipment;
- fire brigade response;
- water supply.

6.4.6 QUANTIFICATION OF RISK

A study of warehouse fires[26] provides the basis for a generic assessment of the frequency of a significant fire. These may be combined with an assessment of the reliability of fire protection systems[27]. Case-specific event trees may be developed from the above data.

The HSE has developed a model known as Firepest[30] which incorporates many of the elements noted above to make an overall assessment of individual risk and in Holland, VROM (the Dutch environment ministry) has undertaken similar studies[31]. Although further developments of this type are likely, at the time of writing they fail to provide the insight of the semi-quantitative approaches. For this reason, care needs to be exercised in developing and interpreting a fully quantitative risk assessment.

6.5 THE ENVIRONMENTAL EFFECTS OF ACCIDENTS

The use of risk assessment for the control of continuous discharges of material into the air or water or onto land is outside the scope of this book. It is, however, appropriate to consider the environmental effects of accidents since the methods used for such assessments have the same basis as those used for safety-related accidents.

Historically, it had been considered that if sufficient measures were taken to prevent major accidents affecting man, then these would also be sufficient to protect the environment. Experience has shown that this is not necessarily the case. Major incidents have occurred which have caused significant environmental damage with little or no damage to man. These incidents have also generated significant adverse publicity. It is, therefore, clear that protection of the environment from accidents may, in certain cases, call for measures of protection which are either different from, or additional to, those needed for the protection of man.

Where spillages of chemicals could lead to the contamination of drinking water, the need for control is clear and regulations are in place in a number of countries, notably Germany. Proposals[32] that would lead to the creation of a water protection zone around the River Dee (which flows through Wales and parts of northern England), are currently the subject of a public enquiry in the UK. If approved, the regulations will require a risk assessment to be carried out in order to gain permission to store defined substances.

6.5.1   HAZARD IDENTIFICATION

In general, the methods used for the identification of safety hazards which are described in Chapter 2, can also be used for the identification of hazards to the environment. Many companies now use integrated procedures to cover safety, health and environmental protection in the design of new facilities[33].

Such combined procedures help ensure that expert staff are used in the most effective way and minimize potential conflict between safety, health and environmental protection.

6.5.2   CONSEQUENCE ASSESSMENT

Where a release is to the atmosphere, the gas dispersion models described in Chapter 3 may be used. For releases to water, a number of models are available. Until recently, relatively simple mixing models have been used. Recent research work has lead to the development of more complex models[34] which take account of mechanisms which can lead to depletion of the contaminant such as volatilization, photolysis, biodegradation, hydrolysis and absorption onto suspended solids.

One of the major problems in carrying out environmental risk assessments is the lack of adequate data on the effect of short-term exposures. Aquatic toxicity data, even where they are available, are mainly determined for the periods of 48 to 96 hours, whereas the duration of an accidental release is typically from 20 minutes up to 1 or 2 hours.

Sometimes data for a similar chemical or species can be used where information is missing. In other cases relationships may be made to long-term environmental quality standards (EQS) with a suggestion that values several orders of magnitude greater than the EQS, could be appropriate for very short duration releases. A further approach under consideration is the establishment of suggested no adverse response limits (SNARLs). Other approaches based on LC_{50} data involve less extrapolation.

Environmental QRAs may be carried out by the methods described in Chapter 4. There is, however, at the time of writing, no clear guidance on criteria of acceptability and tolerability. The US EPA[35] has been using risk assessment techniques for several years to assign regulatory priorities.

The UK Department of the Environment has defined ten classes of environmental damage which could be considered to represent a major environmental incident[36]. But no guidance is provided on tolerable frequencies. Her Majesty's Inspectorate of Pollution (HMIP), in the UK, has established a specialist unit — the Centre for Integrated Environmental Risk Assessment. Further work on this topic is likely.

6.5.3 OTHER WORK

In addition to the methods described already, risk assessment has been applied to soil and groundwater contamination resulting from chemical manufacture, storage or disposal. In one approach[37] risk assessment is used to assess the likelihood of a defined degree of soil and groundwater contamination. In other approaches[38,39] risk assessment is used to assess the likelihood that a measured degree of contamination will spread to reach an environmentally-sensitive site of interest.

6.5.4 DEVELOPMENT

Most development effort has gone into surface water modelling including computer codes such as PRAIRIE (AEA Technology)[34] and VERIS (VROM)[40]. A computerized system (USES — Uniform System for the Evaluation of Substances) has also been developed by VROM to rapidly assess the potential environmental effects of new chemical substances[41]. Although the application of QRA to the environmental effect of accidents is likely to extend, various problems still need to be addressed:
- common treatment of acute and chronic environmental releases;
- establishment of suitable information on the eco-toxicity of chemicals;
- establishment of relevant criteria of tolerability and acceptability.

These factors are likely to limit the application of the techniques in the immediate future.

6.6 SAFETY-CRITICAL COMPUTING SYSTEMS

Modern plant is becoming increasingly complex. New technology allows finer control and thus more efficiently run processes; it also provides novel means of monitoring for increased safety. An important factor in these changes has been the increased use of computer systems, both to control plant and to help ensure its safe operation. Computers can monitor enormous amounts of data and thus allow plant designers to do things that would not be practicable with conventional engineering hardware.

In process plant, computers have been used for several years to help the operators control the plant by presenting them with information in ways that are easily understandable. In some cases they are also being used to control the plant automatically and shut it down safely in an emergency. In many cases computer systems can provide more consistent control and operation than was possible with conventional systems. This is particularly true in the case of batch processing where computer control has produced more consistent products with less likelihood of operator-induced errors. In other cases, such as burner control systems, programmable electronic systems (PES) incorporate additional cross-checking functions which would be difficult to include in conventional systems.

Systems become more complex with the plant foreground control systems being supported by background optimization systems, data loggers, etc, which may link together a large number of separate systems and plants. In addition, modern communication systems make it possible for changes to control variables, or even program elements, to be made remotely, perhaps even from a different country or continent. The operator becomes dependent upon the control system and may lack any 'feel' for how to control the plant in the case of computer failure. Two recent texts have addressed the hazards associated with computer control[42,43].

6.6.1 THE RELIABILITY OF COMPUTER SYSTEMS

This additional complexity of both the plant and the computer systems brings problems as well as benefits; the greater the complexity, the harder it is to be sure that there are no hidden design faults which can reveal themselves as failures during operation. This problem is particularly acute for computer software, which may be unreliable if its designers have not fully understood all aspects of the plant being controlled or if mistakes have been made in writing the program. Some software faults are minor and are relatively easy to detect, rather like spelling mistakes. But even minor faults have the potential to cause serious effects and the more complex the software, the more likely they are to arise and the more difficult to detect.

In the nuclear industry, modern software practices include formal mathematical techniques, which sometimes allow a rigorous proof that a program will behave as specified. Such a specification must itself be a formal mathematical document, however, and there is always a chance that this does not adequately capture all the engineered provisions for the safe functioning of the overall plant. At the present time these techniques are of limited use to other industries.

It is generally not practicable to test a computer program exhaustively for all its different input signals, since the number of different combinations of these is usually astronomical. Fairly modest levels of software reliability can be demonstrated, using a sample of these inputs that is statistically representative of operational use, but they fall short of the levels that are currently demonstrated for comparatively simpler, conventionally-engineered hardware systems.

For all these reasons, there are important limits on the extent to which computer software is relied upon for safety-critical functions. It should be emphasized, however, that these limitations do not preclude the use of computers in safety roles. In the case of a protection system — for example, a primary computer-based system of modest claimed reliability — can be backed up by a simpler, conventionally engineered ('hardwired') secondary system to provide the necessary confidence that shut-down will occur when needed.

Good design practice can assist. Much useful guidance is available in an HSE publication[44]. The quality standard ISO9000–3 is extensively used to control the writing of software, its use being well described in a UK Department of Trade and Industry publication[45].

Where computer systems are used for the control or protection of a plant, one of the hazard identification techniques outlined earlier (for example, Hazop, knowledge-based Hazop and What-if studies) is likely to be used to identify the *process hazards*.

It is strongly recommended that where possible, normal control and optimization functions should be separated from safety-critical functions. This minimizes the chance of a single fault causing both the loss of plant control and the failure of the protective systems.

Where computers are used in safety-critical applications both the hardware and software should be kept as simple as possible, consistent with the system performing its task. The temptation for designers to build programs of ever greater complexity should be resisted. This approach is aided by the separation of control and protective or shut-down functions already noted, since the protective function can often be relatively simple.

Consideration should be given to designing diverse systems to reduce the likelihood of common cause failure. This may be achieved by using different hardware, operating systems and programming languages. The use of hardwired protective systems can maximize diversity and in many cases represents the most cost-effective solution.

There is additional benefit to be gained from carrying out a structured review of the computer hardware, software and support systems. One such approach is the CHazop study[46,47] where a set of guide words and check-lists have been developed specifically for computer installations.

In critical applications fault tree analysis may be applied to assess the reliability of the hardware. This does not represent the total reliability but can often represent a limiting value. The total system cannot be more reliable than the hardware alone.

It is essential that sound procedures are used for recording the basis of the program design. Change control procedures must be introduced and applied throughout the life of the system.

6.6.2 DEVELOPMENT OF STANDARDS

The most important work being undertaken at the time of writing is that of the International Electrotechnical Commission whose standard *IEC 1508 Functional Safety : Safety Related Systems*[48] was issued in April 1995 for consideration and acceptance.

The standard uses an overall safety lifecycle model as its basis. The model incorporates all the activities necessary for ensuring that the required level of safety is achieved for the safety- related systems under consideration. Within the document there is recognition that failures of systems may be caused by either equipment failure or human failure. Human failures can occur during hazard identification, equipment specification, design or operation and maintenance. Requirements are defined for each lifecycle activity; they reduce the probability of systematic error to a level consistent with integrity requirements.

The approach is risk based and can be applied to a range of control and protective systems from purely mechanical protective systems (for example, pressure relief) through conventionally instrumented protective systems to PES-based systems. Detailed guidance has only been prepared for electrical/electronic/programmable electronic systems.

Safety integrity levels are used for specifying the target level of performance for the safety-related systems. The safety integrity level targets are specified taking into account the risk reduction necessary to meet the level of safety specified. Numerical targets for failure measures for electrical/electronic/programmable electronic safety-related systems are linked to the safety integrity levels. Four integrity levels are recognized, giving a range of probability of failure on demand from 10^{-1} to 10^{-5}. The document specifies a limit

Table 6.1
Probability of failure for each safety integrity level.

Safety integrity level	Demand mode of operation (probability of failure to perform its design function on demand	Continuous/high demand mode of operation (probability of a dangerous failure per year)
4	$\geq 10^{-5}$ to $< 10^{-4}$	$\geq 10^{-5}$ to $< 10^{-4}$
3	$\geq 10^{-4}$ to $< 10^{-3}$	$\geq 10^{-4}$ to $< 10^{-3}$
2	$\geq 10^{-3}$ to $< 10^{-2}$	$\geq 10^{-3}$ to $< 10^{-2}$
1	$\geq 10^{-2}$ to $< 10^{-1}$	$\geq 10^{-2}$ to $< 10^{-1}$

in the probability of failure on demand that may be claimed by a single independent system. This has been set at 10^{-5}. The probability of failure for each safety integrity level is shown in Table 6.1.

Although the standard is mainly concerned with safety to man, the approach developed can be applied to environmental hazards and their control as well as to protection against equipment failure and business loss.

Specific guidance on the application of the IEC standard to the process industries is also being developed.

REFERENCES IN CHAPTER 6

1. *Update of loss of containment data for offshore pipelines (PARLOC 92), OTH 93 424*, 1994 (Advanced Mechanics and Engineering).
2. Ramsay, C.G., Bolsover, A.J., Jones, R.K. and Medland, W.G., 1994, Quantitative risk assessment applied to offshore process installations — challenges afyer the Piper Alpha diasaster, *Joural of Loss Prevention in the Process Industries*, 7 (4), 317–330.
3. Morris, M., Miles, A. and Cooper, J., 1994, Quantification of escalation effects in offshore quantitative risk assessment, *Journal of Loss Prevention in the Process Industries*, 7 (4), 337–344.
4. Health and Safety Commission, 1984, *Advisory Committee on Major Hazards, 3rd Report* (HSE Books).
5. Health and Safety Commission Advisory Committee on the Transport of Dangerous Substances, 1991, *Major Hazard Aspects of the Transport of Dangerous Substances* (HSE Books).
6. Saccomanno, F.F. and Cassidy, K. (eds), 1993, *Transportation of Dangerous Goods: Assessing the Risks* (Institute of Risk Research/University of Waterloo Press).
7. Management of process hazards, *API RP 750*, 1990 (American Petroleum Institute).

8. Sweeney, J.C. (ed), 1989, *Guidelines for Technical Management of Chemical Process Safety* (Center for Process Chemical Safety, AIChE).
9. Bond, J., 1988, International safety rating system, *Loss Prevention Bulletin*, 80, 23–29.
10. Reason, J., 1989, Human factors in nuclear power operations in *House of Lords Select Committee on Science and Technology (sub committee II). Research and Development in Nuclear Power, Vol. 2 — Evidence, House of Lords Paper 14.II.* (HMSO) 238–242
11. Health and SafetyExecutive, 1991, *Successful Health and Safety Management HS(G)65* (HSE Books).
12. Bellamy, L.J. and Geyer, T.A.W., 1988, Techniques for assessing the effectiveness of management, *European Safety and Reliability Research Development Association (ESRRDA) Seminar on Human Factors. Bournemouth, March.*
13. Pitblado, R.M., Williams, J.C. and Slater, D.H., 1989, Quantitative assessment of process safety programs, *Plant Operations Progress*, 9 (3), 169.
14. Hurst, N.W., Bellamy, L.J., Geyer, T.A.W. and Astley, J.A., 1991, A classification scheme for pipework failures to include human and sociotechnical errors and their contribution to pipework failure frequencies, *Journal of Hazardous Materials*, 26, 159–186.
15. Hurst, N.W., Bellamy, L.J. and Geyer, T.A.W., 1990, Organisational, management and human factors in quantified risk assessment — a theoretical and empirical basis for modification of risk estimates, *Safety and Reliability in the 90s (SARRS 90)* (Elsevier Applied Science).
16. Taylor, J.R., 1994, *Risk analysis for process plant, pipelines and transport* (Chapman Hall).
17. Ratcliffe, K.B., 1993, STATAS: Development of an HSE audit scheme for loss of containment incidents. Parts 1, 2, 3, *Loss Prevention Bulletin*, Nos 112, 113, and 114.
18. Hurst, N.W. and Ratcliffe, K.B., 1993, Development and application of a structured audit technique for the assessment of safety management systems. (STATAS). *IChemE Symposium Series No. 134*, 315–331.
19. Wackerlig, Hr.U., 1987, Lessons from the Sandoz fire, *Fire East '87 Hong Kong.*
20. CEFIC, 1987, *A Guide to Safe Warehousing for the European Chemical Industry* (CEFIC, Brussels).
21. Cole, S.T. and Wicks, P.H., 1993, *Industrial Fires Workshop Proceedings* (Commission of the European Communities, DG XII, Appeldoorn, The Netherlands).
22. *Fire and its Environmental Impact. A guide to good practice* (Confederation of Fire Protection Associations Europe).
23. *Liste Wassergefährdener Stoffe* (Verband der Chemistre Industrie)
24. *Handbook of Environmental Data on Organic Compounds* (Verschueren).
25. Theobald, C.R, Studies of fires in industrial buildings — part 1: the growth and development of fire, *Fire Prevention Science and Technology*, 17, 4–14.
26. Hymes, I., Flynn, J.F., 1992, *The Probability of Fires in Warehouse and Storage Premises, SRD/HSE R578* (AEA Consultancy Services).

27. Sprinkler focus, 1993, *Fire Prevention*, 257 March, 18–36.
28. Jagger, S.F. and Atkinson, G.T., 1993, The source term for warehouse fires in *CIMAH Safety Reports for Warehouses, 25 March 1993, Chester* (IChemE/W S Atkins).
29. *Fire Protection Concept for Chemical Warehouses with regard to Water Protection*, 1987 (VCI).
30. Bugler J., Kirk P., The effect of fire precautions on the assessment of individual risk from fires in warehouses storing dangerous substances in *CIMAH Safety Reports for Warehouses, 25 March 1993, Chester* (IChemE/ W S Atkins).
31. VROM, *Individual and Group Risk from Storage of Pesticides for Different Fire Fighting Systems, 1990/9* (VROM, The Netherlands). In Dutch.
32. *Proposed Water Protection Zone (River Dee Catchment) Designate Order*, 1994 (National Rivers Authority, Welsh Region).
33. Turney, R.D., 1989, Designing plants for 1990, procedures for the control of safety, health & environmental hazards in the design of chemical plant, *Safety and Loss Prevention in the Chemical, Oil Process Industries, Singapore*.
34. Fryer, L.S., Clark, J.W. and Watson, A., 1993, *Risk Assessment and the Protection of the Environment*, c446/022/93 (Institution of Mechanical Engineers), 113–120.
35. Framework for ecological risk assessment, *EPA/630/R–92/001*, 1992 (US Environmental Prtection Agency, Risk Assessment Forum).
36. Department of the Environment, 1991, *Interpretation of Major Accidents to the Environment for the Purposes of the CIMAH Regulations* (HMSO).
37. Molag, M., 1993, Quantitative environmental of accidental releases of hazardous materials, *Proceedings European Meeting on Chemical Industry and Environment, Girona, Italy, 2-4 June* (Univ Politecnica de Cataluna, Barcelona).
38. *Tackling Contamination. Guidelines for businesses to deal with contaminated land*, 1994 (Confederation of British Industry, London).
39. *Contaminated Land and Land Remediation. Guidance on the Issues and Techniques*, 1993 (Chemical Industries Association, London).
40. VROM, 1994, *VERIS–2 computer code, EnvironmEntal Risk Information System for Seveso Directive Sites* (VROM, The Netherlands).
41. National Institute of Public Helath and Environmental Protection (RIVM), Jager, D.T. and Visser, C.J.M. (eds), 1994, *Uniform System for the Evaluation of Substances (USES), Document 11144/150* (VROM, The Netherlands). RIVM is a division of VROM.
42. Kletz, T., 1995, *Compter Control and Human Error* (IChemE).
43. Health and Safety Executive, 1995, *Out of Control — Why Computer Systems go Wrong and how to Prevent Failure* (HSE Books).
44. Health and Safety Executive, *PES Programmable Electronic Systems in Safety Related Applications* (HSE Books).
45. *TickIT: Guide to Software Quality management System Construction and Certification using EN29001*, 1992 (UK Department of Trade and Industry).
46. Lucas, P.R., 1994, Improving a CHazop process through incident data, *3rd IMECHE Reliability Software Seminar, University of Manchester*.

47. Chung, P. and Broomfield, E., 1995, Hazard and operability (Hazop) studies applied to computer controlled process plants in *Computer Control and Human Error*, Kletz, T. (IChemE), 45–80.
48. *Draft IEC 1508 'Functional Safety : Safety Related Systems*, 1995 (International Electrotechnical Commission).

7. SUMMARY

The first edition of this book commented that risk analysis involves many assumptions on models, data and parameters that detract from the apparent objectivity of the technique. Although these remain, the risk community is emphasizing model validation, release frequency data sources and quality in execution of risk studies. Taken together these should reduce inconsistencies between studies and enhance the value of risk results in routine decision-making.

Other important trends are the extension of the risk approach to cover transport and environmental risks and — at a national level — to assess other man-made hazards than those in the process industry in a consistent manner.

In some parts of Europe, there has been a fundamental shift in regulatory emphasis away from prescriptive rules and towards goal-setting safety objectives that must be adequately demonstrated by operators. Risk assessment techniques are increasingly playing a role in such demonstrations, especially for higher hazard chemical plants or for offshore facilities.

7.1 DEFINITIONS

There now exists substantial agreement on most terms used in risk analysis compared with the situation when the first edition of this book was published. The IChemE has issued guidance accepted widely (Chapter 1, Reference 10). However, new terminology is entering the field, such as ALARP and ALARA.

7.2 HAZARD IDENTIFICATION

(1) It is stressed that hazard identification is an important part of the safety assessment of a plant. The depth of the study and the technique to be used have to be chosen to suit the situation. When the process is concerned with hazardous reactions or toxic materials, the hazard identification must begin at the research bench and continue through the pilot plant or process development stages of a project. Project approval procedures should include the requirement for potential hazard reviews at appropriate stages from the inception of the project,

through project completion and during the life of the operating plant. The type and depth of studies should be determined by the needs at each stage.

(2) Successful hazard identification depends upon having documentation to review which truly reflects the way the plant will be built and operated. The quality of the hazard study is improved by having the designer present his design to the study team.

(3) The methods selected for hazard identification (or the combination of methods) should be those which best fit into the other design and hazard control activities of the particular organization. There is increasing use of formal, structured identification techniques (such as Hazop or equivalent) to document potential hazards and to document the safeguards employed (hardware or procedural). The hazard identification becomes an integral part of the ongoing safety management system of the facility, and is referred to whenever modifications are proposed. Increasingly studies now address safety, occupational hygiene and environmental protection as part of an integrated procedure.

(4) There is growing recognition of the importance of concepts of inherent safety in seeking to eliminate or reduce hazards where this can be achieved cost effectively.

(5) Companies which have used team methods for hazard identification, both for new projects and existing plant, claim that in addition to attaining safer design, operating performance is also improved by the better understanding of operational and maintenance deviations and by motivation of the operating and maintenance staff who have participated in the studies.

(6) Hazard identification procedures are a well-developed element of risk analysis. In the future, greater emphasis on the quality of execution of the identification procedure will be sought. While fundamental changes are unlikely, incremental improvements to enhance team effectiveness and knowledge deployment are currently under development. Some of these systems employ computer-based expert system technology. While this is a promising approach, it is apparent from several European projects that the challenges of the identification task are significant and rapid successes are unlikely. Nevertheless, the role of computer-based expert systems as complementary tools will be an increasing feature of conventional team-based hazard identification.

(7) Where computers are used for the control or protection of a process plant, structured reviews of the computer hardware, software and support systems can be used to improve the overall system reliability. The ISO quality standards (for example, TickIT) and recent developments by the IETC are likely to lead to increasing use of quality standards for control and protective systems, including the associated software, linked to the degree of risk posed by the operation under control.

7.3 CONSEQUENCE ANALYSIS

Substantial progress has been made in the consequence assessment area, with both the development of more mechanistic models and large-scale trials for validation. Commercial software has begun to appear which permits easier access to the technology, although the importance of independent validation is stressed.

(1) Discharge source term calculations have improved substantially since the first edition of this book was published. Fundamental understanding of two-phase release prediction has been gained, with substantial progress on pressure relief problems. Significant advances have occurred in the areas of aerosol formation, droplet rain-out and liquid pool re-evaporation. Given this better understanding, increasing emphasis is being placed on predicting the behaviour of hazardous materials which exhibit non-standard properties, such as hydrogen fluoride.

(2) Considerable work continues in the field of heavy gas dispersion. There are now many available models and a body of large-scale dispersion trial data. A major trend is towards public validation of models for consequence prediction. Other trends will be enhanced model logic for ease of use and automatic linkages to source term models, to allow wider usage of this technology by non-specialists. Methods which model terrain and topography effects are starting to become available.

(3) The phenomenon of vapour cloud explosions and the conditions for transition from a flash fire event have received significant attention. Good tools now exist for offshore module explosion prediction and effect determination, but onshore prediction still lacks this degree of knowledge. Continuing experiments are necessary to establish accurate explosion predictions in real process layouts and neighbouring areas. These will provide an input to revised plant design and layout in order to minimize the possibility of an explosion and its effect on occcupied buildings if there is a release of a flammable vapour or gas.

(4) Prediction of toxic injury in the context of risk analysis is usually based on toxic load or toxic probit approaches. While broadly accepted, there remains considerable uncertainty for many individual gases, including common gases. Toxic impacts are often overpredicted when compared with past experience and this will require greater attention in the modelling of escape and evacuation response. Progress may also come from better modelling of human physiological response to toxic exposure, placing less reliance on animal trials.

(5) A commendable trend during the past several years has been the development of co-operative research projects involving governments, companies and research establishments. This has allowed realistic large-scale trials to be carried out which now permit proper validation of consequence prediction tools.

7.4 QUANTIFICATION OF EVENT PROBABILITIES AND RISKS

There are two broad approaches for event frequency prediction: historical data and system modelling. Both have their place in risk determination.

(1) The historical data approach is the more common method where the objective is whole site risk analysis, and a number of generic data sets have been compiled. While generating reasonable results, these often are based on small data sets or do not have fully traceable origins, and thus may not represent well the particular plant under review. It remains a major challenge for onshore operators to share and pool equipment reliability and release frequency information. In the offshore arena, this data collection and pooling activity has been occurring for some time. Data collection may well proceed independently of QRA, as much of this data is useful in reliability and availability studies. The problem of demonstrating very low probability — that is, obtaining a statistically meaningful estimate for rare events — will always remain. The application of historical data is technically less demanding than the logic diagram approach and can produce more consistent studies.

An increasing trend is the use of modification factors applied to generic historical data to account for site-specific differences. Several projects are underway to account for safety management system quantitative influences and for material damage mechanisms.

(2) The fault tree approach is more commonly applied to detailed analysis of particular items, such as relief systems or reactor controls. Logic diagrams allow a detailed mechanistic failure model to be developed showing the various ways in which a complex system can fail. They allow a thorough understanding of an activity to be built up, enabling persons not familiar with that activity to bring an independent viewpoint to an established procedure or operation. Logic diagrams can be used in both a qualitative and quantitative manner. The qualitative application assists in identification of key areas and safeguards, and provides an aid to communication on how systems may fail and what effect modifications might have. Quantification can provide a clearer indication of the relative importance of the various causes of an undesired event. Quantification also gives a clear view of the relative importance of an undesired event in the overall safety of a particular activity in which a number of such events are possible. Logic diagrams are difficult to construct and require significant expertise and time for complex systems.

(3) Human error is often an important factor in the logic chain which leads to an unwanted event. Identification of the role that human error plays can in itself provide insight into the failure process. Data on the probability of human error quoted in the literature, although based on a number of studies, are still arbitrary and uncertain. Human reliability is expected to remain largely intractable to

quantification except for specific cases. There is some confidence, however, that understanding of the effect of both the operating environment and the basic task structure on the relative likelihood of error in a defined situation will lead to systems which are less prone to human error.

(4) The quantification of event probabilities and risks contains many uncertainties. The quality of data is extremely variable, and errors can be made if the analyst is not fully aware of the data limitations and the theoretical basis of the specific mathematical tools used. An organization proposing to use the techniques must ensure that it commits adequate resources and expertise to the work.

(5) At the time of writing it is not possible to calculate the reliability of computer software with any confidence. There are, however, a number of measures which can be introduced to reduce the probability of software failure.

7.5 THE APPLICATION OF RISK ANALYSIS

(1) In response to some regulatory pressures and to a general enhancement of the technological base, there has been a substantial increase in the use of QRA techniques since the mid-1980s. The experience gained permits a clearer understanding of where such techniques provide value. The areas most suitable for application include comparative analyses of various safeguarding options and general assessment of acceptability for land-use planning purposes.

(2) The comparative mode of risk analysis minimizes uncertainties in the risk results, as it is the difference in risk which is important. Any absolute error tends to be cancelled out. Alternative safeguards can be analysed and their contribution to lower risk through frequency or consequence reduction assessed. Although inherently less uncertain, significant error is still possible and any company applying these techniques must ensure that it has adequate expertise to handle the analytical techniques properly.

Many comparative risk analyses use measures other than fatality as the basis for comparison.

(3) Cultural attitudes and types of legislation are different in the various countries of Europe. A legal requirement to use QRA exists in some countries to demonstrate to the authorities that a plant is adequately safe. In other countries a gradual switch from prescriptive safety legislation towards goal setting legislation increases the uses of QRA. If it is applied as simple absolute pass-fail risk criteria, then serious error may be expected. Calculation errors could lead to rejection of sound proposals and acceptance of unsound ones.

Current land-use planning legislation is increasingly using a three-band risk criterion (intolerable, tolerable if the risk is ALARP and broadly acceptable). While the absolute uncertainty remains, the ALARP zone is defined sufficiently widely that most high hazard plants fall into this zone, and here comparative techniques are used to evaluate different risk-reducing alternatives. Cost-benefit analysis is often used to justify whether or not additional safeguards are necessary.

There are now many examples of application of quantified criteria for determination of acceptability, both in onshore and offshore installations. Early concerns of arbitrary decision making based on unsound risk quantification have not been prominent, and many significant cost savings have been claimed, particularly offshore where some prescriptive safeguards are very costly.

(4) The use of short cut methods for carrying out QRA is not favoured. Nor is the maximum credible accident approach recommended; it can place far too much emphasis on very low probability events. Where a more detailed assessment of a major hazard is required, a QRA should be carried out on the particular part of a plant which is of concern. This should then be a matter for discussion between knowledgeable members of industry and the regulatory body.

(5) Continuing developments in technology are having far-reaching effects on society, including the process industries. These developments will influence the ways of analysing risk, both to people and to the environment in which they live. The future position of QRA depends on how well it can provide information which can be understood and used for decision making within industry and in the public sphere. This could include better ways of dealing with uncertainties and a recognition of the interdependence of reliability studies, QRA and cost-benefit analysis. Regulatory bodies are extending the application of quantified risk techniques into other hazardous human activities, including other industrial operations and all aspects of transportation (with and without the additional hazards of chemicals carried). This is sensible and, over time as consistent risk criteria are applied, this will reduce emotional or unsound decision-making.

(5) Finally it is important to remember that QRA is only one part of the total safety package for the design, construction and operation of process plant. Good engineering, qualified personnel and good management will always be necessary.

INDEX

A
acceptability of risk	90
accident data	72
accidents and 'near misses'; history of	23
accuracy	34, 75, 83, 86
advantages of QRA	88
aerosols	34
ALARA (see as low as reasonably achievable)	
ALARP (see as low as reasonably practicable)	
application of risk assessment	133
in the process industry	87
in the public domain	89
as low as reasonably achievable (ALARA)	91
as low as reasonably practicable (ALARP)	91, 133
audit methods	113, 116

B
BLEVE (see boiling liquid expanding vapour explosions)	
blowdown models	32
boiling liquid expanding vapour explosions (BLEVE)	40, 42
broadly acceptable level of risk	91

C
cause-consequence analysis	71
CHazop	124
check-lists	17
chemical warehouse storage	117
codes of practice	87
coefficient of discharge	33
combustion product effects	117
comparative mode of risk assessment	133
comparison of alternatives	88
compressible flow	32
computer-based expert system	130
computer control and protection	130
consequence	
analysis	29, 131
assessment	118, 120
calculations	87
mode	130
prediction	131
contamination	
firewater	119
groundwater	117, 120
soil	117, 120
cost-benefit analysis	93, 134
cost of a life	93
criteria	
agreed	4
risk	111
tolerability	111

D
damage to humans	54
definitions of terms	7, 129
deflagration	43, 44
dense gases	36
design stage	12
detonation	44
discharge modelling	31

discharge source term calculations 131
dispersion 36

E
effect models 31
elimination of consequences 119
Engineers and Risk Issues, UK
    Engineering Council code
    of practice 6
environmental effects of
    accidents 119
ETA (see event tree analysis)
evaporation of liquids
    on land 35
    on water 35
event probability
    estimation 64
    quantification 133
event tree analysis (ETA) 18, 70, 103
explosion
    damage 49
    prediction models 44

F
failure logic 66
failure modes and effect analysis
    (FMEA) 17
fatal accident rate 79
fault tree analysis (FTA) 18, 67, 132
fireball 40
fire
    damage 55
    protection measures 118
    spread 118
flash fire 39
FMEA (see failure modes and
    effect analysis)
formal, structured identification
    techniques 130
FTA (see fault tree analysis)
fuel-air blast 46

G
Gaussian dispersion model 36
generic
    data 87
    failure rates 114
    frequency 65
guide-word method 14
guide words; lists of 15, 16

H
hazard identification 4, 9, 87, 120, 129
    studies
        composition of the team 20
        organization of the study 19
        preparations for the study 21
        responsibility of the team 19
        techniques 11, 12
            design stage 12
            planning stage 11
hazard indices 11
Hazop 14
    new plants and processes 14
    worksheet 26, 27
heat radiation 40
heavy gas dispersion 37, 131
historical
    data 132
    frequency 65
human error 132
human factors 74

I
identification of hazards 4, 9, 87, 120, 129
IETC (see International
    Electrotechnical Commission)
incident data 72
individual risk 77
inherent safety 3, 4, 11, 130
instantaneous fractional annual
    loss method 115

INDEX

International Electrotechnical
    Commission (IETC)   130
international safety rating
    system (ISRS)   116
International Study Group on Risk
    Analysis (ISGRA)   1
intolerable level of risk   91
ISGRA (see International Study
    Group on Risk Analysis)
ISRS (see international safety
    rating system)

J
jet fires   39, 41

K
knowledge-based Hazop   16

L
land-use planning   133
legislation and QRA   133
limitations of QRA   86
logic diagrams   66

M
management assessment
    guidelines in the evaluation of
    risk (MANAGER)   115
material hazard assessment   117
maximum credible accident
    approach   134
minimal set cut analysis   68
multi-energy method   46

N
neutral gas regimes   36
new plants and processes
    and Hazop   14

O
offshore QRA   98

P
PES (see programmable
    electronic systems)
physical effects   31
planning
    enquiries   89
    permission   95
    stage   11
PLL (see potential loss of life)
pool fire   40
potential loss of life (PLL)   93
probability data   71
probit equations   60
process drawings   22
professional responsibility   6
programmable electronic
    systems (PES)   122
project procedures   129

Q
QRA (see quantified risk assessment
    and risk assessment)
quantification of event
    probabilities   133
quantified risk assessment (QRA)
    advantages   88
    legislation   133
    limitations   86
    links with safety management
        systems   115
    offshore   98
quantitative expressions of risk   77

R
rain-out   34
releases to water   120
reliability and event data   72

reliability of computer software/systems	122, 133
research	131
responsibilities	93
risk	
calculation	81
control	111
criteria	111
individual	77
quantitative expressions of	77
terminology	77
risk assessment (see also quantified risk assessment)	
application	133
in the process industry	87
in the public domain	89
special topics	98
what it is	4

S

safety	
cases	100
critical computing systems	122
integrity levels	124
management	94
systems	113
Seveso Directive	89, 117
short cut methods	134

smoke	
hazards	42
product effects	117
societal risk	79, 95
sociotechnical audit method	115
sources of knowledge	22
structural damage	49

T

task analysis	18
thermal	
fluxes	41
radiation	39
three-zone approach	91, 111
TNT-equivalence methods	45
tolerability of risk	90
toxic injury	57, 131
transport risks	108
turbulence	43

V

validation	39, 48
vapour cloud explosions	43, 131
vulnerability models	49

W

What-if/check-list methods	17